Human Heat Stress

Human Heat Stress

Ken Parsons

CRC Press is an imprint of the
Taylor & Francis Group, an informa business

CRC Press
Taylor & Francis Group
6000 Broken Sound Parkway NW, Suite 300
Boca Raton, FL 33487-2742

Printed on acid-free paper

International Standard Book Number-13: 978-0-367-00233-6 (Hardback)

Library of Congress Cataloging-in-Publication Data

Names: Parsons, K. C. (Kenneth C.), 1953- author.
Title: Human heat stress / Ken Parsons.
Description: Boca Raton, FL : CRC Press/Taylor & Francis Group, 2019. |
Includes bibliographical references and index.
Identifiers: LCCN 2018050354 | ISBN 9780367002336 (hardback : acid-free paper) |
ISBN 9780429020834 (ebook)
Subjects: LCSH: Heat exhaustion. | Heat--Physiological effect.
Classification: LCC RC87.1 .P36 2019 | DDC 616.9/89--dc23
LC record available at https://lccn.loc.gov/2018050354

Visit the Taylor & Francis Web site at
http://www.taylorandfrancis.com

and the CRC Press Web site at
http://www.crcpress.com

To Jane, Ben, Anna, Hannah, Richard, Sam,
Mina, Nancy, Thomas and Edward

Contents

Preface

From the time I started writing this book in late spring of 2018 to submission in the fall, there have been hundreds of heat-related deaths in North America and thousands of casualties reporting to hospital caused by hot weather. This is repeated across the northern hemisphere (summer) after many heat casualties had been reported over the previous summer in the southern hemisphere. In Japan, a national disaster was declared in July when 30 people died and 30,000 casualties reported to hospital in 1 week. In the United Kingdom, where temperatures rose to over 35°C, there were over 1,000 excess deaths when compared with previous years.

The most frequently referred to catastrophe due to heat was in Paris in 2003, where there were over 14,000 excess deaths, mainly among the elderly, and thousands admitted to hospitals, in a 2-week period, leading the responsible government minister to resign. Newspapers are not always a source of reliable data, but they provide evidence of what seems important both now and in the past with reports and images ranging from entertaining to alarming. In 1911, the UK Daily Mail showed an image of young ladies keeping cool during a heat wave by paddling in the Serpentine in Hyde Park, London, fully clothed, bonnets and all. In the same article, 772 deaths are reported in 1 week among children from London and Manchester, mostly related to problems of heat and hygiene.

If we restrict our commentary to mainly the summer of 2018 and UK newspapers, we see celebrities featured eating ice cream, using fans and the Dalai Lama addressing an audience with a wet towel on his head. Many people are shown lying on beaches in swimwear, and children around the world are shown sitting in large buckets of water, under waterfalls and playing in public water features of various types. A tiger is seen licking a large ice block, a panda in China is lying on one, and a giraffe in England is seen licking a giant fruity "ice lolly." Soldiers in London, England and Arlington, USA are shown collapsed on parade. There are deaths from children and animals being left in hot vehicles, and in sport, one person died and there were hundreds of heat casualties in the London Marathon (24.1°C), where on average one person dies every three years. A world champion tri-athlete helped his brother, who was suffering from heat exhaustion, across the line in Mexico. A super fit champion tennis player lost his match in the U.S. open tournament, complaining of breathlessness due to heat, and in Australia, tennis players and supporters are shown cooling in giant fans with added sprayed water. At the Wimbledon tennis tournament, male players complained that female players were given heat breaks but not men, and at the same tournament, 22 supporters needed medical treatment as a heat stress index rose above 30.1°C.

An extensive study by Harvard University suggested that for every 0.5°C rise above comfortable temperatures, there is a loss of 1% in examination performance among children who became more likely to be distracted and agitated in the heat. As well as the crisis in Japan, where temperatures were above 40°C, the highest ever recorded temperatures in Africa were reported (Algeria, 51.3°C) and 70 people died due to heat stress in Quebec, Canada. In the United Kingdom, headlines reported that Britain was in meltdown, medical facilities were closed down due to heat, hospitals were in chaos, rail lines were buckling, workers were on 'go slow' (a formally agreed reduction in workrate), and on Thursday July 26, 2018, we were told to brace ourselves for "Furnace Friday," where temperatures reached 37°C. The aforementioned is only a snapshot, and all in one summer.

In September, there was a court martial trial of two soldiers who were responsible for risk assessments for a military exercise where three highly motivated aspiring "special forces" soldiers died of heatstroke during an exercise. The soldiers were acquitted because it was considered that they had received insufficient training. In Denmark, on 8 August, a report showed an increase of about 250 deaths, primarily among the elderly, in the summer of 2018, when compared to the norm. A similar increase was reported in the neighboring countries affected by the heat wave, including other Scandinavian countries.

The reports above are mainly from UK newspaper articles and may vary in their accuracy of actual numbers of those affected by heat including deaths, heat casualties, or extreme distress and discomfort. What is without dispute is that heat affects health, causes injury and death, and takes over the news agenda. Everyone is affected by it and interested in how to keep cool. The brief summary mentioned above can be repeated worldwide and requires serious consideration. It is impressive how once the weather cools interest vanishes. A consideration of the nature of heat stress, how it affects people, and how to avoid heat casualties is timely, and despite the repeated consequences for health and general alarm, we should be encouraged by how much we know.

This book provides a modern, up-to-date, concise, and comprehensive coverage of the fundamentals behind the assessment of heat stress and how they can be applied to avoid heat casualties.

Chapter 1 introduces what we mean by heat stress and its relevance in context throughout the world. It describes the basic factors which are internationally accepted as the variables that are most important and essential as a starting point when considering human response to hot environments. Chapter 2 considers human response to heat and presents the importance of homeostasis and the system of human thermoregulation. It is uniquely presented in its modern form, which includes both the physiological and the behavioral responses of people, acting in unison in an attempt to prevent unacceptable thermal strain and ensure survival. Chapter 3 considers the important personal factor of metabolic heat production which is related to activity level, and Chapter 4 considers heat stress and clothing which greatly influence the ability of the body to lose heat in hot conditions. Methods of measuring the relevant aspects of the environment that make up heat stress are presented in Chapter 5. These include air temperature, radiant temperature, air velocity, and humidity. Methods of measuring thermal strain, including body temperature, heart rate, skin temperature, and sweating are presented in Chapter 6. Chapters 7–9 consider how to assess hot conditions to predict how humans will respond and provide requirements for avoiding heat casualties. These include the wet-bulb globe temperature index in Chapter 7 and methods based upon the body heat balance equation and the calculation of required sweat rate in Chapter 8. Both are presented with respect to the most up-to-date international standards. Chapter 9 presents computer models of human thermoregulation that can be used to predict human response to the heat.

Chapter 10 presents heat illnesses and how to prevent them, and Chapter 11 considers human performance and productivity in the heat. Chapter 12 presents burns and thresholds for skin and surface temperatures that would cause a burn when in contact with solid surfaces. Up-to-date knowledge and references, including those derived from outcomes of projects on global warming, universal climate indices, and protective clothing and others, from consortia of international scientists, are provided along with an appendix which provides useful websites to go to for further information.

Ken Parsons
October 2018,
Loughborough

Author

Ken Parsons is an emeritus professor of environmental ergonomics at Loughborough University. He has spent over 30 years of research and application into human response to heat. He was born on January 20, 1953, in northeastern England in a coastal village called Seaton Sluice. He graduated from Loughborough University in ergonomics in 1974, obtained a postgraduate certificate in education in mathematics with a distinction from Hughes Hall, Cambridge University in 1975, and was awarded a PhD in human response to vibration in 1980, from the Institute of Sound and Vibration Research, Southampton University. He founded the Human Thermal Environments Laboratory at Loughborough in 1981 and was awarded a certificate in management from the Open University in 1993. Ken became the head of the Department of Human Sciences in 1996, Dean of Science in 2003, and pro-vice chancellor for research from 2009 to 2012. He was the chair of the United Kingdom Deans of Science from 2008 to 2010.

In 1992, he received the Ralph G. Nevins Award from the American Society of Heating, Refrigerating, and Air-Conditioning Engineers (ASHRAE) for "significant accomplishments in the study of bioenvironmental engineering and its impact on human comfort and health." The Human Thermal Environments laboratory was awarded the President's Medal of the Ergonomics Society in 2001. He is one of the co-authors of the British Occupational Hygiene Society publication on thermal environments and has contributed to the Chartered Institute of Building Services Engineers publications on thermal comfort as well as to the *ASHRAE Handbook: Fundamentals.*

He has been a fellow of the Institute of Ergonomics and Human Factors, the International Ergonomics Association, and the Royal Society of Medicine. He was a registered European ergonomist and an elected member to the council of the Ergonomics Society. He has been a scientific advisor to the Defence Evaluation Research Agency and the Defence Clothing and Textile Agency and a member of the Defence Scientific Advisory Committee. He has been both the secretary and the chair of the thermal factors committee of the International

Commission on Occupational Health (ICOH), chair of the Centre National de la Recherche Scientifique (CNRS) advisory committee to the Laboratoire de Physiologie et Psychologie Environnmentales in Strasbourg, France, and is a life member of the Indian Ergonomics Society. He was a visiting professor to Chalmers University in Sweden and is a member of the committee of the International Conference on Environmental Ergonomics. He was an advisor to the World Health Organization on heat waves and a visiting professor to Chongqing University in China, where he was a leading academic to the National Centre for International Research of Low Carbon and Green Buildings. He was the scientific editor and co-editor in chief of the journal *Applied Ergonomics* for 33 years and is on the editorial boards of the journals *Industrial Health, Annals of Occupational Hygiene,* and *Physiological Anthropology.*

He is co-founder of the United Kingdom Indoor Environments Group and a founding member of the UK Clothing Science Group, the European Society for Protective Clothing, the Network for Comfort and Energy Use in Buildings, and the thermal factors scientific committee of the ICOH. He was the chair of ISO TC 159 SC5 "Ergonomics of the Physical Environment" for over 20 years and is convenor to the ISO working group on integrated environments, chair of the British Standards Institution committee on the ergonomics of the physical environment, and convenor of CEN TC 122 WG11, which is the European standards committee concerned with the ergonomics of the physical environment.

1 Human Heat Stress

HUMAN HEAT STRESS

All people at all times respond to thermal environments, and many people throughout history and across the world have been incapacitated, injured, or have died of heat stress and more continue to do so. As a consequence, there has been much research into how people respond to heat, when heat casualties will occur, and how to avoid heat casualties. Fundamental knowledge has led to an understanding of heat stress and universal methods for considering heat stress and the thermal strain it leads to in people. We use this knowledge to quantify heat stress and predict likely thermal strain.

Heat stress is often considered to be caused mainly by high temperatures, but it is fundamental to an understanding of heat stress and its consequences that it is seen to be a combination of four environmental variables (air temperature, radiant temperature, air velocity, and humidity) and two personal factors (clothing and metabolic rate). It is essential in any consideration of the effects of heat stress on people that all of these six factors are considered.

A model of the universe is that it is made up of energy that can neither be created nor destroyed. So in a finite, albeit expanding, universe, there must be a finite amount. The energy is distributed across the universe from areas of no energy (not quite achieved in practice), where everything is still, to varying levels of energy, which makes things move (or change state, liquid to gas, for example). The more energy within an object, generally, the greater the movement of its component particles. The average amount of energy due to motion (kinetic energy) in an object is termed its temperature. When things are absolutely still, this is regarded as an absolute temperature of zero. If all of the energy were concentrated in one place, the temperature would be extremely high and there may be a "big bang."

Although energy concentration varies across the universe, energy flows only in an irreversible direction from hot to cold, from areas of high concentration to those of low concentration, and in general, from order to disorder (entropy). Energy has a natural tendency to dissipate, so that that over time (which moves forward but not backward) it will eventually become evenly spread across the universe and stabilize at a uniform temperature. The consequences for the earth will be that it will cool down. This is our long-term destiny. At least, that is how we see it at present. To understand how people respond to heat and determine methods of assessment of human heat stress, it is important to recognize that people are objects in the universe and subject to its conditions.

THE HUMAN DISPOSITION

In the meantime, thousands of people die and many more experience discomfort, physiological and psychological strain, and illness due to exposure to energy in the form of heat. Exposure to heat from outside of the body is termed heat stress. The human body also produces metabolic heat inside of the body. When combined, the total heat can produce a reaction of the person, and this is called heat strain.

The human disposition is to maintain in the body a level of heat energy that is required to maintain every person at an internal temperature of around 37 degrees Celsius (°C. [Under normal atmospheric pressure, 0°C is the temperature of melting ice and 100°C is the temperature of boiling water. Absolute temperature is –273.15°C and on the absolute temperature scale is 0 Kelvin (K)]. Note that 1 K = 1°C temperature difference. Water boils at 212 degrees Fahrenheit (°F) and freezes at 32°F. This provides 180° between the two. To convert °F to °C take off 32, multiply by 5, and divide by 9. To convert °C to °F divide by 5, multiply by 9, and add 32.

The heat to maintain body temperature is generated in the cells of the body by burning food in oxygen. Food comes from plants using energy from the sun (or via other animals), and oxygen is in the air that we breathe. This is called metabolic heat, which increases with our level of activity.

If the resting human body were completely insulated (encapsulated in thick clothing, blankets, and so forth), then no metabolic heat would transfer out of the body, the heat would accumulate, and body temperature would rise at about 1°C h^{-1}. If the body is to be maintained at around 37°C, then the metabolic heat needs to be transferred to the environment. If an egg is cracked into water at, at least, 44°C, then the clear albumin will solidify and turn white. The protein has denatured in an irreversible reaction. Human tissue will also denature at, and above, 44°C, so this provides an extreme upper limit for human survival in the heat and demonstrates the terrible consequences of exceeding that limit. This value of 44°C is far too high for an acceptable safety limit for the body. A value of 38°C–39°C may be more appropriate. How long it will take to reach these limits, may be regarded as safe exposure times.

Heat (energy) can be measured in joules. One calorie is 4.186 J and is the heat required to raise 1 g of water by 1°C. Power is the rate of production or transfer of energy and is measured in watts, where 1 W is equal to $1\,J\,s^{-1}$. If a recumbent resting person of mass 70 kg produces 80 W of metabolic heat and the average specific heat capacity of a person is $3.49\,kJ\,kg^{-1}\,K^{-1}$, then a 1°C rise to 38°C would require 3.49×10^3 (convert to joules) $J\,kg^{-1}\,K^{-1} \times 70\,kg$. That is 244,300 J of heat at $80\,J\,s^{-1}$ (W). It would therefore take an insulated human body around 51 min to reach 38°C and therefore 102 min to reach 39°C. If we take body temperature as a criterion for safety limits, then we can calculate "safe" exposure times.

The principles described previously will be the same if the person is not completely insulated by "clothing" but where the environment is so hot that no heat can transfer out of the body (i.e., insulted by the environment). Some environments may even increase the heat content of the body and some, which is the "normal" case, will allow heat to transfer from the body to the environment, usually involving clothing which allows some restriction but also some heat transfer.

If we consider a more usual situation where a person is not recumbent but sitting or standing at rest or even more active, maybe wearing clothing, and that the environment is not so severe such that it does allow heat to transfer from the body to the environment, then a thermal audit (Parsons, 1992, 2014) can be performed. In effect, this is a method of accounting for, and quantifying, all the avenues of heat to the body and lost from the body. If net heat lost is equal to net heat gained, then the body temperature will be stabilized, and if there is more heat in than heat out, the body temperature will rise. It is worth reflecting at this stage that the laws of thermodynamics apply to all things in the universe, and that includes people. Energy can neither be created nor destroyed, so we can account for the heat and that the heat transfers only from a hot human body to an environment at lower temperature or a hot environment to a cooler human body. It also applies to the transfer of vapor from areas of high vapor concentration, such as those caused by evaporated sweat next to hot skin, to areas of low vapor concentration, such as in lower temperature and humidity environments.

THE BODY HEAT EQUATION

For an object to maintain a constant temperature, it must maintain heat energy at a constant level. The human body is in continuous and dynamic interaction with its continuously changing and dynamic environment. Within this variation, it "attempts" to maintain a "constant" internal temperature of around 37°C. It does this by continuously gaining or losing heat, to and from the environment, so that the net gain or loss is zero. The net result is that when this is achieved it is termed "heat balance." The mechanism for achieving heat balance is described in the heat balance equation.

For an object to maintain constant temperature, the net heat transfer (sum of gains and losses) into and out of the object by dry heat loss (conduction, convection, and radiation) and latent and vapor transfer (evaporation and condensation) will be zero. That is, if we quantify the heat transfer by those avenues and sum the values (positive and negative), if the total is zero, there will be no heat gain and the temperature will remain constant. This applies to the human body with the addition that the total must include the metabolic heat generated by the body.

The heat balance equation for the human body is therefore a heat gain due to metabolic heat on one side of the equation, which equals (is balanced by) the sum of heat gains and heat losses by conduction, convection, radiation, evaporation, and condensation. In general, conduction is usually regarded as negligible and ignored (except for local effects such as contact with hot surfaces causing pain and burns), and condensation is regarded as the heat gain equivalent of evaporation. In terms of the body heat equation, heat storage (positive or negative) is added for cases where insufficient heat is lost to the environment (positive heat storage and body temperature rise) or too much heat is lost to the environment (negative heat storage and fall in body temperature).

The usual form of considering the human heat balance equation is to consider metabolic heat as the total of metabolic free energy production (M—metabolic rate) in the body minus any energy that is required for mechanical work (W). It cannot all be heat, as we need some energy to move around. For most activities, most of the energy is produced as heat and the mechanical energy is relatively small (usually <10% of metabolic rate and not easy to estimate accurately). This is balanced by the heat lost (as a sum of losses and gains) by convection, radiation, and evaporation (of sweat) from the skin as well as heat lost through breathing (transfer of dry air and water vapor). The heat balance equation introduces the concept of what is required for heat balance, such as how much sweating is required and hence provides an indication of the likely thermal strain on the body.

THE SIX BASIC PARAMETERS

To represent an environment in terms of factors that will influence heat stress and determine thermal strain, a minimum of six factors should be considered. Actually, these are variables as they all vary with time, but it is usual to consider them as parameters that represent their values at any point in time. These are the air temperature, radiant temperature, humidity, and air velocity, which represent the environment and the personal factors of clothing and metabolic rate. It is the interaction of the factors, and not any one or sub-combination of factors, that determine any human response. We can determine the relevance of these factors by considering the human body heat equation in more detail. Figure 1.1 presents the body heat balance equation and the avenues of heat transfer between an active person and the environment.

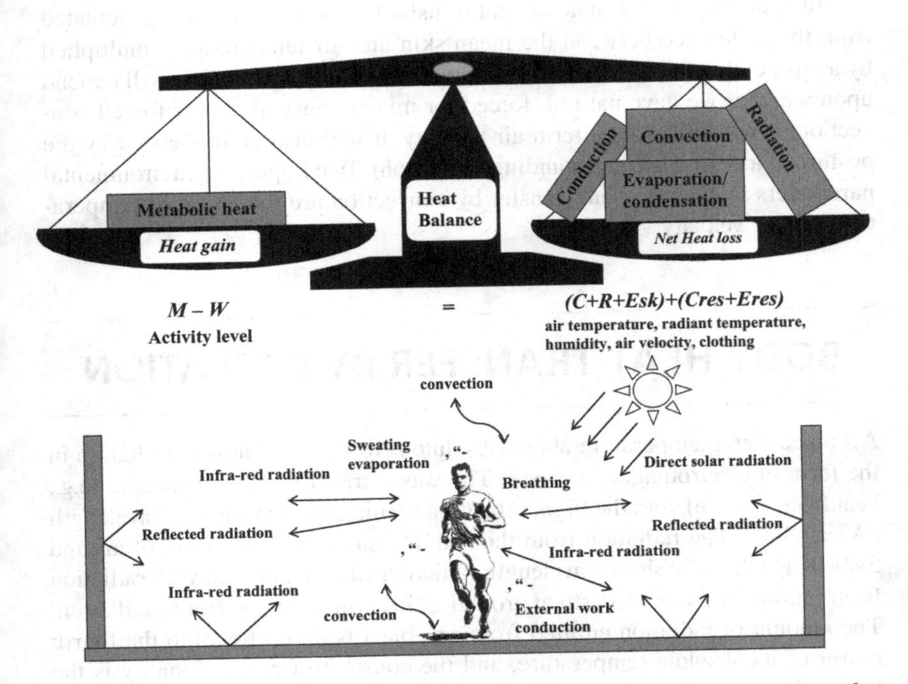

FIGURE 1.1 The body heat balance equation and avenues of heat transfer between an active person and the environment.

BODY HEAT TRANSFER BY CONVECTION

Heat transfer by convection is due to the movement of fluid, and the heat is carried by the fluid itself. In the case of the human body, it is the movement of air (or other medium such as when in water) across the surface of the body. The driving force for the heat transfer is the difference between the skin temperature (or clothing surface temperature) and the air. If the body is hotter (at a higher temperature) than the air, heat is lost from the body. If the air is at a higher temperature than the body, heat will be gained to the body by convection. If the body and air are at the same temperature, then there is no net heat transfer by convection. If there is no wind, the hotter body will heat cooler air, which will rise with more dense cooler air replacing it due to gravity. This is called natural convection. No gravity, no natural convection, hence, this does not occur in space. If there is relative movement between the body and the air (e.g., caused by body or air movement or both), then the fluid (air) is replaced by forced convection.

In summary, an estimate of heat transfer by convection can be calculated from the difference between the mean skin and air temperatures, multiplied by a term called the convective heat transfer coefficient (hc). This will depend upon whether we have natural, forced, or mixed convection. For forced convection, it will include the term air velocity. It will also be influenced by the posture of a person (sitting, standing, and so on). Two important environmental parameters relevant to heat transfer by convection are therefore air temperature and air velocity.

BODY HEAT TRANSFER BY RADIATION

All objects at a temperature above absolute zero emit and absorb radiation in the form of electromagnetic waves. The waves travel through a vacuum (e.g., heat from the sun), and the higher the temperature, the shorter the wavelength (Wien's law). The radiation from the sun at a surface temperature of around 5,800 K is relatively short wavelength radiation when compared with radiation from indoor surfaces on earth at around 297 K which emit infrared radiation. The amount of radiation emitted from an object is proportional to the fourth power of its absolute temperature, and the constant of proportionality is the Stefan–Boltzmann constant ($\sigma = 5.67 \times 10^{-8}\,\mathrm{W\,m^{-2}\,K^{-4}}$). The heat from the sun arrives at the surface of the earth's atmosphere at an intensity of around 1,370 $W m^{-2}$. After passing through the atmosphere, this provides a maximum of around 1,000 $W m^{-2}$ at the earth's surface in full sun.

Radiant heat is exchanged between all objects, so a person will receive radiation from all surrounding surfaces and emit radiation to all surrounding surfaces. The net radiation "transfer" is proportional to the difference in the fourth powers of the absolute temperatures of all surfaces as they interact with each other. There will be a net input of radiant heat from surfaces at higher temperature to surfaces of lower temperature (e.g., from the human body to cooler surfaces in offices or to the human body from industrial furnaces, radiant heaters, or the sun). This difference is often taken as the difference between the mean skin temperature (or mean surface temperature of clothing) for a person and the mean radiant temperature as a three-dimensional average of all the surrounding surface temperatures in an enclosure. That is a representative of, and often derived from, the temperature at the center of a black sphere. If a surface is black, then it emits and absorbs all wavelengths of radiation. It has an emissivity (ε) of 1. For a nonblack surface, in particular, a white or silvered surface, the emissivity may be much less than 1, but that depends upon the wavelength of radiation. For short wavelength radiations, the emissivity will vary with the color of the surface, but for relatively low temperature surfaces, emissivity will approach that of a black body ($\varepsilon = 1$) irrespective of the color. For that reason, wearing white clothing outdoors in the sun will have lower heat absorption to a person than wearing darker-colored clothing. However, wearing white clothing indoors out of the sun will have no advantage in reducing heat to the person over darker clothing.

A further factor relevant to heat transfer in solar radiation, or where most of the radiation comes to the body from one direction (e.g., from a steel furnace or the sun through a window), is the amount of surface area of the body that is facing the sun (hot surface). Using photographic techniques, Underwood and Ward (1966), have provided radiation surface areas for different postures and orientations of the body with respect to the sun.

In summary, an estimate of heat transfer by radiation (*R*), can be calculated from the difference between the fourth powers of the absolute temperatures of mean skin (or clothing) temperature and the mean radiant temperature for the shape of the person at the position of the person. This is multiplied by a term called the (human) heat transfer coefficient for radiation (hr) which is a combination of the emissivity, projected radiation surface area, and the Stefan–Boltzmann constant ($5.67 \times 10^{-8} W m^{-2} K^{-4}$). An important environmental parameter, essential for assessing heat stress, is therefore the mean radiant temperature.

BODY HEAT TRANSFER BY EVAPORATION

The latent heat of vaporization is the heat required to change liquid to vapor without a change in temperature. Evaporation will "convert" the water from liquid sweat to vapor at the surface of the skin, and the heat required will be taken from the skin surface and hence cool the person. Water vapor can be contained in air without influence on its other constituent parts. The air adjacent to the skin, at skin temperature, will increase in (water) vapor concentration as evaporation occurs, until it can hold no more vapor. This is a state of saturation. If the water vapor concentration in the air in the environment outside of the body is lower than the concentration of water vapor at the skin, then (if clothing allows) vapor (at skin temperature) will transfer from the higher to lower concentration and heat will be lost from the body to the environment as air at the skin is replaced by air from the environment to complete the cycle. The water vapor concentration in air exerts a "water vapor pressure." This "partial pressure" in the air is usually used in the calculations of heat transfer. The driving force for heat transfer is usually taken as the difference between the saturated vapor pressure at skin temperature and the vapor pressure in the surrounding environment which is the partial vapor pressure at air temperature. Factors that affect heat transfer by evaporation are similar to those that affect heat transfer by convection. Although not always stated, there is, by analogy, a natural evaporation and a forced evaporation where air flow forced (by wind) across the skin will increase the rate of evaporation and hence heat transfer. For this reason, it is convenient to use the Lewis relation (LR) (Lewis, 1922), which states that the ratio of the heat transfer coefficient by evaporation (he) and the heat transfer coefficient by convection (hc) is constant (LR = 16.5 K kPa^{-1}).

It is important to note that just as clothing insulation and dynamics (ventilation, pumping of air inside the clothing, and so on) are important when considering dry heat transfer ($C + R$) to and from the body, the vapor permeation and dynamic properties of clothing will be important for heat transfer by evaporation.

In summary, an estimate of the heat transfer by evaporation from the body can be calculated from the difference between the partial vapor pressure at the skin (saturated vapor pressure at skin temperature) and the partial vapor pressure in the air in the environment. This is multiplied by an adjustment for properties of clothing and the (human) heat transfer coefficient by evaporation (he) which is directly related to the heat transfer coefficient by convection (hc) by the LR. Humidity, which is directly related to partial vapor pressure in the air, is therefore an important parameter when considering heat stress.

Returning to the six basic parameters essential for heat stress assessment, we see from the heat transfer analysis that air temperature, radiant temperature, air velocity, and humidity all influence heat transfer between the human body and the environment. In addition to these four environmental parameters, two personal factors are metabolic heat production and clothing. The metabolic rate provides heat generated inside of the body and is related to the level of a person's activity, which also influences their rate of breathing. Clothing properties influence the transfer of dry heat, water, and water vapor between the body and the environment. All six parameters are important and provide a minimum starting point for heat transfer assessment.

REGULATING BODY TEMPERATURE AND COMFORT—HUMAN THERMOREGULATION

When people become hot, they feel it mainly through sensors in their skin, and they take any available opportunities to behave in a way which avoids or reduces the heat stress. This is behavioral thermoregulation, often termed an adaptive approach to thermoregulation, and the opportunities to behave appropriately to avoid heat stress are called adaptive opportunities. Examples include moving out of the sun to shade, reducing clothing levels, switching on fans, and many more. Behavioral thermoregulation can be very powerful and successful if adaptive opportunities are available or are made available by environmental design. Although often a very effective response, behavioral thermoregulation is a mainly conscious response where a threshold of unacceptability or dissatisfaction is reached (e.g., feeling too hot).

There is also a continuous system of thermoregulation that controls a subconscious or automatic physiological response. This involves detecting the condition of the body, mainly in terms of temperature and rate of change of temperature and an integration of signals from across the body to a central processing system where a decision is made regarding an appropriate response. This response is in terms of effector mechanisms to ensure that the state of the body is optimum. In other words, orientating the body towards losing heat if the body is too hot and towards preserving or generating heat if the body is too cold.

The detection of the thermal state of the body is via nerve endings in the skin, some specifically for heat and the others for cold. They respond to the temperature of the skin as well as to the rate of change of temperature. Their signals are integrated with signals from sensors inside of the body to present the brain (hypothalamus) with a representation of the thermal condition of the body. This is compared with a desirable or preferred condition, and a response is provided in proportion to the difference between the perceived actual, and desired physiological, condition. For example, if the internal body temperature is significantly greater than a desired 37°C, then effectors are enabled to stimulate heat loss.

In the first instance of heat stress, there will be a tendency for the body temperature to rise and blood is directed from the center of the body to the skin by vasodilation. This raises the skin temperature and increases the potential for heat loss by convection and radiation. There is a continuous feedback loop as the sensors monitor the effectiveness of any action. If insufficient heat is lost due to this vasomotor control alone, then sweating promotes heat loss by evaporation.

Physiological thermoregulation is an automatic, continuous feedback system that mainly involves conditioning the skin to lose the required amount of heat in hot conditions. It works in conjunction with the behavioral thermoregulation where a person can feel the state of the skin through temperature and mechanical sensors to detect discomfort when the skin is wet with sweat.

FEELING HOT

When people are exposed to heat stress, they have a feeling of being hot, and they can report that feeling subjectively. Subjective scales of stickiness are often used to describe discomfort when hot due to the interaction of sweating skin and clothing. There are other suggested scales for use in hot environments; however, subjective scales should be used with great caution and never as safety criteria to allow people to continue to be exposed to heat. There are two important factors that mitigate against using subjective scales to assess people in extreme environments. The first is that the effect of the environment on a person may cloud his or her judgment. A person may not be able to make a valid judgment of whether he or she is in danger, or not, even if they have the skill to do so. The second is that there are contextual factors that will influence the perception and reporting of how safe a person may consider an environment to be. Motivation provides a major risk factor. Highly motivated people, in sporting competitions, military exercises, when paid for productivity and more, will overstate their ability to survive under heat stress. This is combined with a feeling of satisfaction and achievement due to what has been achieved and what is perceived can be achieved. The feeling of exertion and extreme condition of the body (profuse sweating) when exercising in hot conditions provides a feeling of euphoria rather than discomfort. For this reason, under no circumstances should a person who is exposed to heat be the judge of whether it is safe for them to continue to be exposed to the heat. Their judgment to terminate their exposure must be accepted, but their judgment to carry on must be ignored. A judgment by a trained person involving the person's behavior, disposition, rationality, and demeanor, as well as their measured physiological state, must be made. There have been many deaths of motivated people who have insisted that they are fit to continue with a heat exposure.

A more detailed discussion of thermoregulation and human responses to heat is provided in Chapter 2.

Thermoregulation and Human Response to Heat 2

THE HOT ENVIRONMENT

On a cloudless sunny day in June 1961, in the famous scientific outdoor laboratories at Rothamsted in the United Kingdom, at 52° north on the Greenwich meridian (0° W), a pyranometer measured solar radiation at midday at around 850 W m^{-2} (Monteith and Unsworth, 1990). This is a similar value that Simon Hodder and I measured outdoors near Seville in Spain (mean afternoon values over 8 days of 891 W m^{-2} direct, 93 W m^{-2} diffuse, 154 W m^{-2} albedo, with a maximum direct solar radiation value of 983 W m^{-2}). Measurements were made during an investigation into human response to heat in vehicles by driving them and recording the responses of the passengers, between Seville and Cadiz, a hot part of Europe with a consistent clear sky (Hodder and Parsons, 2001; Hodder, 2002). A similar maximum value (895 W m^{-2}) was also recorded in the summer of 2008 when Ju Youn Kwon and I measured climatic conditions and human responses across a whole year in a weather station in Loughborough, UK (Kwon and Parsons, 2009; Kwon, 2009). The sun at a surface temperature of around 5,500°C emits around 3.88×10^{26} W, and a portion of this travels the 93 million miles (circa 150 million km) to the earth. The intensity of this direct solar radiation has been measured by satellites at the surface of the atmosphere as 1,373 W m^{-2} (the solar constant), the exact value depending upon the orbit of the earth around the sun. Direct solar radiation at the surface of the earth is therefore at a maximum of around 1,000 W m^{-2} depending upon how thin the atmosphere is at that point and how much heat is scattered and absorbed. At night, the earth must lose this heat so that it can maintain a heat balance and "control" average temperature of around 14.6°C. The energy from the sun is distributed around the earth by the weather. Temperatures have ranged from a recorded maximum of 56.7°C (134.1°F) in Death Valley in the United States to –89.2°C (–128.6°F) in the Russian, Soviet Vostok Antarctic research station. Wind velocities range from still air to 253 mph (408 km h^{-1}; 113 ms^{-1}) recorded in Tropical Cyclone Olivia on Barrow Island, Australia. Surface winds had been recorded at 231 mph on Mount Washington, New Hampshire, United States (speeds in a hurricane have been recorded up to 190 mph and in a tornado up to 301 mph). Relative humidity ranges in hot conditions from close to 0%, for example, in the desert, to 100%, for example, in the rain forests (0 g m^{-3} of dry air to around 51 g m^{-3} of dry air at 40°C) when precipitation may occur causing rain. Energy levels flow from hot to cold and from high concentration to low concentration, the earth spins and is tilted, and the moon (and sun) causes tides as gravity moves the seas and oceans.

This is the environment on earth on which organisms survive. One strategy for survival in this varied environment is for an organism to maintain a relatively constant internal environment where it carries out essential functions despite variation in the external environment. It usually does this by varying the condition of the outside of its body to maintain conditions relatively constant inside. This is called homeostasis, and was described by Claude Bernard in his work on what he called the "internal milieu." All mammals, including humans, practice homeostasis, and in terms of maintaining a relatively constant internal temperature, they are called homeotherms. People "attempt" to maintain an internal body temperature of around 37°C and usually within the range 36.5°C–37.5°C. To achieve this, all people have a system of thermoregulation.

HUMAN THERMOREGULATION

Heat stress occurs when the environment creates conditions where the internal body temperature of a person has a tendency to rise. Generally, the human response to the heat stress is to attempt to increase heat loss from the body to the environment or reduce heat gain to the body from the environment. This thermoregulation occurs as two types. One is mainly conscious, where a person will respond to discomfort or dissatisfaction by taking action (e.g., moving away, adjusting clothing, opening a window, or switching on a fan). This is called behavioral thermoregulation or an adaptive approach. The second is an unconscious, automatic, and continuous system called physiological thermoregulation. The systems work together and are shown in Figure 2.1.

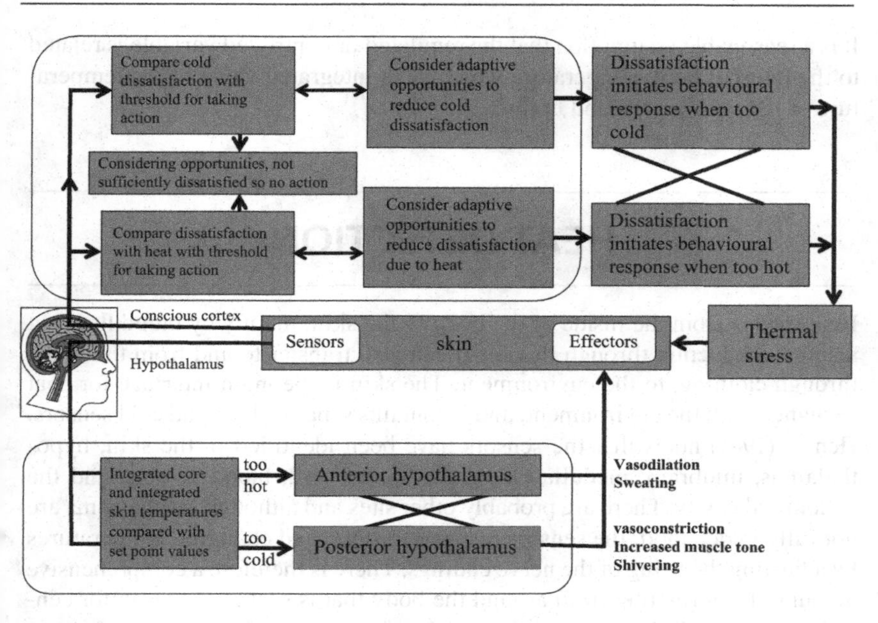

FIGURE 2.1 A diagrammatic representation of human thermoregulation. Behavioral regulation is controlled by the conscious cortex and mostly requires a conscious decision. Physiological regulation is continuous and automatic and controlled by the hypothalamus.

PHYSIOLOGICAL THERMOREGULATION

The physiological system of thermoregulation can be conveniently divided into two parts, the controlled system and the controlling system. The controlled system is often called the passive system as it represents the nonactive part of the human body. That is the skin, fat, muscle, bone, organs, head, hands, feet, arms, legs, trunk, lungs, blood, lymph, and so on. The dimensions, shapes, and thermal properties of the components of the passive system will be important in thermoregulation. The controlling system controls the state of the passive or controlled system in an "attempt" to ensure that the internal body temperature can remain at around 37°C through loss of excess metabolic heat and gain from the environment. The controlling system therefore has a system of detection and transmission, a system of integration and processing, and a system of effecting physiological response (e.g., vasodilation and sweating).

It is a reasonable assumption that the regulated or controlled variable is related to the internal body temperature which is an integrated value of the temperature of the vital organs and brain.

HEAT DETECTION

Heat travels from the inside of the body to the skin, mainly by blood flow but also by conduction through tissues. It can also transfer to and from the body, through clothing, to the environment. The skin is the main interface for heat exchange with the environment, and it contains separate heat and cold sensors. Hensel (1981) notes that the sensors have been identified in the skin, hypothalamus, midbrain, medulla oblongata, spinal cord, blood vessels, and the abdominal cavity. There are probably other sites and although mechanisms are not fully understood, the sensors respond to static and changing temperatures by adjusting the firing of the nerve endings. There is therefore a comprehensive amount of information from around the body that is sent to the brain for central processing. It should also be noted that there are local responses to heat so there can be some thermoregulation near to the position of the heat stimulus (e.g., if you heat the skin locally, it vasodilates and sweats).

The response of the skin sensors is not linearly related to temperature, and the rate of firing of nerve endings has a different temperature response depending upon whether they are hot or cold sensors. The hot sensors are deeper in the dermis of the skin and fire at a higher temperature range than the cold sensors. When outside of the range of effective operation, the sensors provide anomalous results leading to extreme cold being reported as hot for example. Adaptation also occurs where sensations reduce in intensity over time even though the physical stimulus does not change. A party trick is to place the left hand in cold water and the right hand in warm water. After a time, place both hands in "neutral" water. The left hand will feel warm and the right hand will feel cold even though they are in the same water and receive identical stimuli. Kenshalo (1970) has demonstrated this more systematically for forearm temperatures, for both the adaptive and the rate of change of temperatures. If the rate of change of temperature is sufficiently slow, (e.g., $<0.02\,K\,s^{-1}$) then, from a neutral starting point, people cannot detect a change in temperature of 1°C cooler or up to 3°C warmer. This is part of the rationale behind the method of cooling people in water baths in heat waves without causing a thermal shock (Parsons, 2014). The temperature sensors in the body respond to temperature, but their response is nonlinear and the information transferred to the brain requires integration and interpretation.

CENTRAL PROCESSING

Edholm and Weiner (1981) note that, from the results of studies using electrical stimulation, the thermosensors from around the body send signals to two linked hypothalamus centers. The hypothalamus is situated just behind the eyes (preoptic region). The warm (hot) sensors to the anterior hypothalamus which controls heat loss due to vasodilation and sweating, and the cold sensors to the posterior hypothalamus which controls the preservation of heat by vasoconstriction, heat generation by shivering, and piloerection. We should note that heat transfer is also by breathing but that in humans, unlike in other animals such as dogs, panting is not a thermoregulatory response. By a mechanism not fully understood, the brain compares its perceived state, derived from information from sensors, with a desired state that seems to be "preprogramed." This is to achieve and maintain a temperature of around 37°C, but will vary with the time of day, the time in a monthly cycle, and other factors such as rate of exercise and fever.

The details of the response mechanisms are not fully understood, but the effector mechanisms are stimulated at a strength that increases in relation to the difference between what is optimum and what is considered to be the actual physiological condition of the body. There will be a threshold involved (however narrow) where no action is taken, and beyond that threshold, effector mechanisms are stimulated. There are many possibilities for control systems. Hensel (1981) points out that technological solutions which are usually proposed are only a small subset of all possibilities, particularly when considering organic systems. Internal body temperature is often regarded as the primary controlled variable; however, whatever is controlled or regulated is clearly related to the strength of the integrated (electrical and chemical) signals from the body to the brain. The response can involve a varying threshold, absolute or relative differences, and rate of change (or higher order) of variables with time.

Three concepts are often used, none of which have physiological counterparts. These are body core, which represents the internal part of the body whose temperature is controlled; the body shell, which varies in temperature and condition and is controlled to ensure that heat transfer is what is required; and the set point (which is neither set nor a point), which is the desired core temperature. The set point will vary with the time of day, time of the month, work rate, and with other factors, such as fever. We can consider the set point as the point at which the body reacts in an "attempt" to return to optimum conditions. Interestingly, this increases as the activity level increases (Nielsen, 1938; Leithead and Lind, 1964; Nielsen and Nielsen, 1965). In this case, the core temperature that is "defended" increases with work rate. There is some debate about whether the set point actually changes. Sawka and Wenger (1988) note that during exercise, the core temperature is proportional to the metabolic rate. They conclude that this is independent of the environmental conditions, however, and that it is due to a "load error" and not a change in set point. The rise in core temperature with activity level occurs in the prescriptive zone (Lind, 1963; WHO, 1969) where heat balance is achieved mainly by appropriate levels of sweating. The term prescriptive is used as it is in the range of conditions where safe industrial conditions can be prescribed. The upper limit of the prescriptive zone (ULPZ) is the highest level of heat stress that the body can endure before internal body temperature begins to rise. A rise in internal body temperature would indicate heat storage; hence, heat stress levels above the ULPZ are said to be in the environmentally driven zone. In summary, set point temperatures can be based upon core temperature, skin temperature, mean body temperature, rates of change of temperature, and more. The responses are proportional in a graded way to the difference between the set points and the value of the actual "reported" controlled variable in the body (Sawka and Wenger, 1988).

HUMAN PHYSIOLOGICAL RESPONSE

The physiological "effector" system used to effect the control of body temperature involves stimulating blood flow to the skin and sweating. These are the main mechanisms for influencing heat transfer between the body (through clothing) and the environment.

BLOOD FLOW

Blood flow influences the temperature of the skin, and sweating influences how wet it is. Blood perfuses through the whole body and generally takes on the temperature of the surrounding tissues. As it moves around the body, it therefore transports heat. If 1 L of blood transfers from the body core at 37°C to the skin, and returns from the skin at 36°C, then 3.85 kJ of heat will be lost (Sawka and Wenger, 1988). The distribution of blood around the body is determined by signals to blood vessels from the control system and by local effects. The rate of blood flow is determined by the heart. This includes heart rate and stroke volume. During exercise, muscles heat up and the heat is transported from the muscles by blood. When a person is experiencing heat stress, vasodilation allows more blood to flow to the skin surface. An increase in skin temperature increases heat loss (or reduces heat gain) by convection and radiation. It is particularly effective in the limbs (arms, hands, fingers, legs, feet, and toes) as they have a large surface area-to-mass (volume) ratio. In hot conditions, countercurrent heat exchange is avoided. This is where cool venous blood runs next to warm arterial blood, transferring and transporting heat back to the body. Blood also flows to the skin surface, avoiding circumstances which direct arterial blood directly to venules via small blood vessels called anastomoses. These mechanisms are used to preserve body heat when cold and are therefore not employed when hot.

During exercise, there is a competition for blood between exercising muscles for metabolism, and the skin for heat loss. There is a central allocation system, which, as well as blood flow itself, is enhanced by local control.

SWEATING

The first significant response of the body to heat stress is to increase skin temperature using blood flow. If sufficient heat is lost using this mechanism, then this "vasocontrol" provides thermoregulation. If insufficient heat is lost then sweating occurs. The body is covered with sweat glands (estimated up to 4 million of them) each of which is a small winding tube in the dermis with an opening at the skin. Sweat is a combination of water and various electrolytes and minerals. It is produced within the tube and transported to the skin surface. If 1 L of sweat evaporates from the skin surface at 36°C, then approximately 2,430 kJ of heat is lost from the body to the environment. Evaporation of sweat from the body is a powerful method of survival during heat stress and is the only effective "physiological" mechanism of heat loss when environmental temperatures are approaching and surpassing body temperature. In fact, the skin is never completely dry even under conditions of no heat stress. The so-called insensible perspiration, from the skin, will provide around 5–10 W of heat loss from a person by evaporation depending upon environmental conditions and clothing. (There will also be a small amount of heat transfer by vapor transfer during breathing in addition to the heat loss by convection during breathing.)

A fit acclimatized person with a close to maximum level of exercise of around 800 W can potentially evaporate up to 2–3 L h^{-1} of sweat to lose that excess heat to the environment. If 1 g of sweat at 30°C evaporates, then 2.43 kJ of heat is potentially removed from the body. If the sweat drips, then it is lost without significant heat loss. Assuming all sweat evaporates at the skin surface, then $2\,L\,h^{-1}$ (2,000 g) provides a heat loss of $2{,}000 \times 2.43 = 4{,}860\,kJ\,h^{-1}$. An activity level of 800 W is $0.8\,kJ\,s^{-1}$ so $0.8 \times 3{,}600 = 2{,}880\,kJ\,h^{-1}$. If we take account of the inefficient dripping and maybe restrictions due to clothing, the person may just about survive for a short time at that level of exercise in the heat (if dehydration is also considered).

Sweat glands are termed eccrine glands, and they are distributed unevenly across the skin and are activated in an uneven way as heat stress increases. The highest density of sweat glands is found on the souls of the feet and the lowest on the back (Sawka and Wenger, 1988). These are not necessarily the primary areas for evaporation, however, as sweat flows out of the sweat glands and runs over the skin surface. Sweat glands vary in size between individuals so "sweaty" people often have larger sweat glands.

As heat stress increases and there is a need for sweating, the sweat glands are increasingly recruited by region and sweat per gland increases. Smith and Havenith (2011) have produced body maps of sweating athletes. Nadel et al. (1973) present the density of sweat glands for individual regions across the body and note that recruitment starts in the back and chest and moves to the limbs only after a further increase in core temperature. They report that at a constant skin temperature, sweat rate is proportional to the core temperature; at a constant core temperature, sweat rate is proportional to the skin temperature; and at combined core and skin temperatures, local temperatures enhance the sweat rate.

ACCLIMATIZATION

Central and local stimuli cause sweating as part of human thermoregulation in the heat. An enhanced part of the system is that sweat rates can more than double if people become acclimatized to the heat. For example, a fit man could raise his maximum sweat rate from $1.5\,L\,h^{-1}$ to up to $3\,L\,h^{-1}$ for short periods, with associated advantage to heat loss but with additional hydration requirements.

Acclimatization is a process where the body and the person "learns" to reduce thermal strain for a given level of heat stress. This occurs through physiological change to allow heat loss from the body to become more efficient and effective (e.g., an increase in sweating capacity) and a behavioral change where people learn how to reduce, avoid, and cope with the heat stress. This part is therefore context and culturally related. The term "acclimation" is essentially acclimatization when conducted in the laboratory or under controlled conditions. In fact, acclimatization initially would have been used as a general description of getting used to the climate in the long term (Kenney et al., 2012). It is possible to acclimatize (acclimate) someone by exposing them to heat stress, severe enough to raise body core temperature and promote sweating (and evaporation of sweat) for a few hours per day over days. This is best done by exercising people in the heat but can also be achieved by passive heating of a resting person in a ventilated suit, as long as the body temperature rises and the sweat evaporates (Fox et al., 1964). The physiological results achieved in acclimation are identical to those of acclimatization. In acclimation, acclimatization is achieved in the context of the laboratory as opposed to other contexts, so it is debatable whether the term "acclimation" is required. Behavioral acclimatization will depend upon context and, therefore, will also apply to the laboratory, although in a laboratory context, almost by definition, there will be limited adaptive opportunity when compared with other less controlled contexts.

Programs of acclimation have been shown to reduce heat casualties (Wyndham and Strydom, 1969; Wenger, 1988) and are recommended when people are exposed to heat stress. For example, when starting a program of work in the heat or returning to hot work, from a weekend or vacation. Military personnel, who are to be deployed to a hot region, or individual or team sports men and women, who are going to compete in hot climates, will particularly benefit, although it is a great advantage if the person who is being acclimated is already at a high level of fitness.

For workplaces and industries who do not have access to a laboratory, it is useful to define acclimatization as having worked in the hot environment or its equivalent for at least 1 working week. This is useful as heat stress limits can be raised if people are acclimatized to heat so a practical working definition is required (ISO 7243, 2017).

Exercise in the heat for 1 h or more in hot conditions by unacclimatized people will show a significant change within a few days and be completed over a week or two (Ladell, 1964). For a given level of exercise and heat stress, heart rate will reduce and sweating begins earlier as it begins at a lower level of core temperature than when unacclimatized. Kenney et al. (2012) describe the sweat and blood flow changes with acclimatization but also note an increase in blood plasma volume over the first few days which they say facilitates other changes. Wenger (1988) presents the results of three male subjects exercising on a treadmill in hot and cool (25°C, control) conditions. On the first day, heart rate and rectal temperature rose to well above those experiencing cool conditions. After 10 days, average heart rate reduced by 40 bpm and rectal temperature by 1°C, that is, to values similar to those in the cool conditions. In addition, sweat rate rose by 10%, skin temperature fell by around 1.5°C, and there was a small drop in metabolic rate. In addition, sweat becomes more dilute as there is a reduced loss of electrolytes.

Lind and Bass (1963) found similar responses after a 9-day period of exercising in the heat. Sweat loss increased rapidly from day 1 to day 5 and then increased more gradually up to day 9. At the same time, rectal temperature and pulse rate dropped in a similar way as presumably the increased heat loss by evaporation, due to increased sweating, had reduced the thermal strain.

There is some debate about the rate of loss of acclimatization. It is agreed, however, that there is some loss after a few days and that it will be significantly reduced after a few weeks (Kerslake, 1972). Wenger (1988) notes that, during acclimatisation, lower thresholds for sweating and skin blood flow reflect a shift in set point temperatures. These are accompanied by a local increase in sensitivity of the sweat glands and increased capacity as well as an increased ability to reabsorb salt.

It can be concluded that a substantial part and pivotal role in physiological acclimatization is the "training" of the sweat glands (Collins et al., 1965; Kerslake, 1972; Wenger, 1988; Parsons, 2014): that is, an increase in the capacity of the glands to produce sweat by exercising the sweat glands by exposing them to heat. If sweating is restricted or inhibited during the heat exposure, then acclimatization does not occur. To allow a continuous flow of sweat, evaporation must occur, otherwise the glands become blocked. This can be caused by sweat and water vapor trapped beneath clothing and hence blocking evaporation (hidromeiosis). It emphasizes the importance of using naked skin, high air velocities, and ventilated suits to achieve acclimatization. For a person wearing a fully encapsulated impermeable suit (e.g., nuclear, biological, and chemical suits often worn by military personnel), there may be little advantage in acclimatization and some disadvantage due to increased sweating that drips causing water loss but no increase in heat loss apart from some loss by condensation at the clothing surface and heat transfer by conduction through the clothing (Lotens, 1993).

Turk (1974) described a method used by the British Army to acclimatize large groups of soldiers. They performed step exercises for 1 h/day over 4 days, in large hot enclosures and self-regulated their oral temperature at around 38.5°C. A similar procedure had been used previously by Wyndham and Strydom (1969) to reduce heat casualties and deaths in the South African Gold mines. It is important to note that some people are heat intolerant (NIOSH, 2016). A priority should be given to identify the heat intolerant as they will be the most vulnerable when considering casualties and heat stress. It is also generally agreed that fitness is an advantage for achieving acclimatization as well as for heat tolerance. It is often confounded with motivation, however. Parsons (2014) uses the term "sensible" to include behavioral acclimatization (self-activated appropriate action) and the implementation of appropriate management procedures. His advice for not being "shafted by the heat" is SHAFTS (Sensible, Hydrated, Acclimatized, Fit, Thin, and Sober). Sober includes the avoidance of the use of drugs, including alcohol, which may affect thermoregulation and hydration as well as behavior and judgment.

Clark and Edholm (1985) report studies that investigated whether people who normally live in a hot climate differ in their acclimatization process. They found that when soldiers from India (a hot climate) were acclimatized after spending a winter in the United Kingdom, they showed no difference in the acclimatization status and process than that of the UK soldiers. They also found that sweat rates increased in proportion so that those with highest (lowest) sweat rates before acclimatization were those with highest (lowest) and increased (more than doubled) sweat rates after acclimatization. That is the rank order of those that sweated most to least did not change with acclimatization. This is consistent with acclimatization involving the training of the sweat glands.

BEHAVIORAL ACCLIMATIZATION

It can be seen in Figure 2.1 that the system of human thermoregulation has two parts that run together. Physiological thermoregulation is automatic, subconscious, and continuous, and its controlled variable is mainly core temperature or something related to it. Behavioral thermoregulation is mostly conscious, more threshold based, and hence manifestly discreet. It is reasonable to hypothesize that any behavioral response is related to the degree of unacceptability or dissatisfaction with the current position. This may be due to the level of discomfort, for example, and any behavioral response will be driven by dissatisfaction and a threshold for taking action, taking account of the kind of action available (adaptive opportunities). A person may be hot, sticky, and dissatisfied with the thermal environment, for example, but in a formal interview may not be willing to take off clothing or move out of the room.

Fanger (1970) links discomfort under steady state conditions to dissatisfaction. He identified three factors for thermal comfort. These are internal body temperature, skin temperature, and sweat rate. We can add local thermal discomfort due to draughts, temperature gradients, and asymmetric and high levels of thermal radiation (e.g., the sun). The regulated variable, in behavioral thermoregulation, is therefore related to discomfort or dissatisfaction and the homeostatic position is to maintain satisfaction and comfort. Behavior can therefore be interpreted as reducing dissatisfaction, and it is achieved by reducing the heat stress to the body. This can be viewed as altering the combination of air temperature, radiant temperature, air velocity, humidity, clothing, and activity to which the body is exposed.

A simple method of reducing heat stress is to adjust or remove clothing. Another is to move to a less hot environment. Technical solutions can involve switching on air conditioning or increasing air velocity using fans. There are many examples and more to be discovered. Providing the opportunity to use them is a challenge for the environmental designer. Some are obvious and others are learnt through experience and cultural practices across the world.

An important factor is what behaviors are available in any context. This is called adaptive opportunity. Designers can use methods that complement the use of the space such as the use of a "Marilyn Monroe" area or blasting and extracting cold air across dancers in a night club. A simple example, of restricted adaptive opportunity, is where a strict dress code is required such as soldiers marching in a royal or presidential parade. No opportunity is available for moving away from the heat stress, adjusting clothing, or even changing workload (a paced task). People performing agricultural work in fields in the sun may seem to have an opportunity to slow down (self-paced) or move to shade, but if they are paid by productivity (e.g., amount of fruit picked), the motivation and necessity to earn, counteract adaptive opportunity. An important area is where wearing protective clothing and equipment as protection from a hazard (e.g., a toxic environment) reduces or negates any adaptive opportunity. In this case, heat stress can be severe and dangerous especially where the clothing cannot be removed easily.

The principles of adaptive, or behavioral, thermoregulation are shown in Figure 2.1. Adaptive opportunity is an integral and essential part of modern presentations of human thermoregulation and responses to heat stress.

A final note concerns whether different populations, within and between cultures and countries throughout the world, differ in their responses to heat stress. There have been a number of studies (Clark and Edholm, 1985); however, there is no evidence that the principles of thermoregulation shown in Figure 2.1 do not apply to all people. There is little evidence to suggest differences in physiological responses; however, it is natural that cultural and contextual differences affect adaptive opportunity and specific behavioral response. There are populations of people with disabilities, illnesses, and so on. The specific disability or illness will affect both physiological and behavioral thermoregulation. The principles shown in Figure 2.1 will still apply, but the specific responses, particularly those of adaptive opportunity, will be important. For example, reduced sweating capacity, lack of mobility, or ability to remove clothing will greatly increase the risk of injury in hot conditions (Parsons, 2014; Kenny, 2017; Inoue and Kondo, 2017).

Human Metabolic Heat

3

BODY HEAT

Dead people do not produce heat. In fact, we can estimate the time of death of a person from their body temperature and the thermal conditions in which their body has lain. Living people take in air and extract oxygen from it, which is transported by blood to all living cells around the body. They also take in food and water, break it down, and transport it in the blood to the cells around the body. Inside the cells, structures called mitochondria convert a chemical called adenosine diphosphate (ADP) to adenosine triphosphate (ATP) using the energy from the food (now glucose) and oxygen in a chemical process called Kreb's cycle. This is called respiration. The energy is therefore stored for "later" use in the phosphate bond of ATP. This is a simplified description. For more detail, see Kenney et al. (2012). We should mention that this can occur without oxygen for a short period (anaerobic respiration), carrying an oxygen debt which is repaid later. There are many complicated nuances and pathways to the above, but the main outcome is that we can store energy in the body on the phosphate bond on ATP and release the energy, forming ADP, hence a cycle. Much of the energy is as heat, which is transported around the body mainly by blood to maintain our homeothermic requirement of an internal body temperature of 37°C and releasing any excess to the environment. Some of the energy is used to activate muscles and perform other body functions. The total (*Meta*-bolism) amount of energy produced, integrated across all cells, is called metabolic rate (*M*—sometimes called the metabolic free energy production). The total heat produced across all of the cells is termed metabolic heat production (*H*), and the energy used for moving the body and other bodily functions is called mechanical work (W) such that $M = H + W$ or $H = M - W$. We should clarify a common confusion. W here is nothing to do with the efficiency of a machine a person is operating (e.g., a bicycle). It is exclusively to do with the operation of the muscles and other bodily functions. If the machine is efficient, then the work done by the body is put to good use but W is to do with the body not the machine. For most tasks, W is small (<10% of M) when compared with H. In practice, it is difficult to determine and measure metabolic rate with accuracy, and as W is small, H is often taken as M (i.e., $W = 0$). It is often said that the body is inefficient in energy utilization as W is much less than M. However, H is not wasted energy and is an integral part of the human disposition as a homeotherm.

UNITS

The units of energy are joules (J) and the rate of production of energy are watts ($W = J\,s^{-1}$). So metabolic rate is measured in watts. Here we need to clarify the confusion as units are often reported that attempt to standardize across people. The clarification is that the metabolic rate is the integral of all the energy produced in the cells, the more cells, for the same level of activity, the more the energy. This is our starting definition. In an attempt to standardize metabolic rate values for a given activity, so that a single value for an activity applies to all people, we need a standard way of reporting for all human bodies. As all people are different, this is impossible to achieve to a high level of accuracy, if at all so, any method will be an approximation. We may say that "larger" people have more cells hence, will produce more energy for a given activity than smaller people. However, we must also consider that some cell types produce much more energy than others (e.g., muscles) and that this will depend upon the activity type and level. Thus, the somatotype (fat, muscular, or thin) of the person will be important.

Two of the most commonly used standardization methods are to use body mass and body surface area. We can say that all people throughout the world sitting at rest produce 58 W for every square meter of their body surface area (actually the unit 1 Met = $58.15\,W\,m^{-2} = 50\,kcal\,m^2h^{-1}$). Those with larger surface areas will produce more heat and those with a smaller surface area will produce less, but both will produce $58\,W\,m^{-2}$ sitting at rest. An argument for using body surface area ($1.8\,m^2$ is often used as an average for men and $1.6\,m^2$ for women, although in modern times and across all age groups, 2 and $1.8\,m^2$ may be more accurate due to secular trends and obesity) is that it is related to body size (number of cells) and is usefully related to the size of the primary surface for heat exchange (the skin). Body surface area is often estimated from the mass (weight, Wt, kg) and height (Ht, m) of a person using an approximate equation (DuBois and Dubois, 1916) and called the DuBois surface area ($A_D = 0.202\,Wt^{0.425}\,Ht^{0.725}$). Using mass to standardize metabolic rate values (J $kg^{-1}s^{-1}$; W kg^{-1}; and kcal $kg^{-1}h^{-1}$) may be a better representation of heat produced in some activities, walking uphill, for example. For tasks involving moving body weight, it may be more valid to differentiate people, and hence, standardize units, based upon body mass. Clearly, a heavy person will use much more energy than a thin person of the same height, when walking up stairs, and the heat produced, integrated over all cells, in particular the muscles, is better represented in this way. Although, it should be noted that the body surface area is influenced by, and estimated from an equation involving, body mass. There are other measures that can be used to standardize units across people, such as body volume, muscle mass, and so on. The conclusion, however, is that each individual is unique as is his or her metabolic rate (in watts) for a given activity. In practice, and in analysis, however, it is convenient to use standardized metabolic rate values so that approximate values can be provided and used that depend upon activity type but not the specific individual or group.

MEASUREMENT AND ESTIMATION OF METABOLIC RATE

Whole Body Calorimetry

If we totally immersed a person in an insulated water bath, say in 1,000 kg (L) of stirred water at 30°C with surrounding air at 30°C, with appropriate mechanisms for breathing, and we observe the water temperature rises to 30.2°C in 2 h, what does that tell us? If the heat to raise 1 kg of water at 30°C by 1°C is 1 kcal (4.186 kJ) then the human body has heated the 1,000 kg of water using 837,200 J, in 2 h, for a 0.2°C rise ($0.2 \times 4.186 \times 10^3 \times 1{,}000$). There are 3,600 s in an hour so 837,200 divided by $2 \times 3{,}600 = 116$ W. For stirred water, the skin temperature will be fixed at water temperature (increasing from 30°C to 30.2°C) and we assume that core temperature can be maintained at 37°C. Thus, we can assume that all of the metabolic heat is transferred to the water. So, metabolic rate for the person can be estimated as 116 W.

Even with this simple example of whole-body calorimetry, we can see how complication can enter the procedure. Placing someone in water while breathing underwater, through a tube, is probably stressful. If we add to this the hydrostatic pressure of the water, change in breathing patterns due to air resistance and stress, and change in body weight and abnormal posture, it can be seen that it would be difficult to gain a valid measure of heat production during activity using this very simple calorimeter. We have calculated an approximate value and can make assumptions and refinements, but it is not perfect. For the reasons given previously, water bath methods do not provide useful results when attempting to estimate metabolic rate for different activities. There are a range of water bath and water suit methods that have been used for this purpose (NIOSH, 2016) but none overcome the problems described.

DuBois (1937) proposed a controlled room where avenues of heat exchange were monitored between inside the room containing the person and outside of the room. By accounting for heat and mass transfer through different avenues, an estimate of metabolic heat production could be made. Murgatroyd et al. (1993) describe a gradient layer chamber method, where heat losses are measured passively across the walls of a poorly insulated chamber, and heat sink calorimetry, where heat is extracted from a well-insulated chamber. Evaporation is measured by an increase in air humidity. Modern whole body calorimetry can provide accurate measures of net heat coming out of a person but provide difficulties if used outside of the laboratory. For practical applications and typical activities, these methods are of limited value.

Indirect Calorimetry

Whole body calorimetry estimates the metabolic heat production by attempting to account for all the heat produced. Indirect calorimetry estimates metabolic rate from the resources used to produce it. That is, food and oxygen. A bomb calorimeter burns a fixed amount of food (any combustible material) in oxygen and measures the heat produced. For 1 g of food, we find this produces 15.76 kJ for pure carbohydrate, 39.4 kJ for pure fat, and 18.55 kJ for pure protein (Murgatroyd et al., 1993). For a "normal" mixed diet, a value of 5.68 Wh (20.45 kJ) per litre of oxygen is often used to estimate the energy produced by a person.

When we burn a carbon-based substance, such as food in oxygen, the carbon combines with the oxygen to produce carbon dioxide. The efficiency of the combustion is related to how much carbon dioxide is produced, when compared with how much oxygen is used to burn the substance. When this is considered for respiration, it is called the respiratory quotient (RQ). For carbohydrate, the RQ is 1.0; for protein, the RQ is 0.835; and for fat, the RQ is 0.710. For a mixed diet, the RQ is often taken as 0.85.

So, if we measure the amount of oxygen used by the body over a fixed period and the amount of carbon dioxide produced (giving the RQ value), we can estimate the energy produced by the body and hence the metabolic rate. So how do we do this?

There are a number of systems. These include the K-M system, which is a device, such as a square box worn on the back, full-face mask breath-by-breath systems and ventilation systems (see NIOSH, 2016). Douglas (1911), working in the famous respiratory laboratories of Haldane and others, invented the Douglas bag. This was a canvas wedge-shaped gas bag (twill, lined with vulcanized rubber) of about 1 m^2 and volume of 50 L. The Douglas bag is supported and used to collect expired air from a person performing an activity using a nose clip (not to be forgotten) and a three-way mouthpiece (breath in external air and expire to the outside or breath in external air and expire to the bag), with a wide connecting tube and valve to open and shut the bag. The subject should be allowed to settle into the activity by performing it without the apparatus for a while and then wearing the mouthpiece, bag, and so on, but breathing in external air and expiring to outside. The bag is suspended at head height. All effort should be made to ensure that the activity is as representative as possible to that of interest, and when ready, the bag is opened and the mouthpiece valve directs expired air into the bag. After a fixed time (no less than 5 min—more time the lower the level of activity and hence breathing) and preferably when the bag is close to full but not straining, the bag is closed. The ensemble is not pleasant to wear. When the bag is closed, the mouthpiece, nose clip, and accompanying saliva are swiftly removed into a bucket of water (I let the subject do this) providing a face wipe to the subject to clean up. For multiple collections, to measure metabolic rate for a series of activities, for example, a rotating washing line system of bags, hanging from the struts of the umbrella-like system, can be used. The bags should be at head height with short tubes between mouth and bag to reduce air resistance.

There are many potential problems, and training of the person making the measurement is required. The stress of wearing the mouthpiece and the nose clip can influence breathing (hyperventilation). The bags may diffuse gas and analysis of the expired air should be conducted soon after collection. For fuller description of the procedure see Douglas (1911), Bonjer et al. (1981), Murgatroyd et al. (1993), and Parsons (2014).

The volume of, and percentage of oxygen in, expired air, its temperature at that volume, and the percentage of carbon dioxide in expired air (to calculate RQ) are all measured using calibrated instruments. Standard gases are often used for calibration. Assuming 20% oxygen in inspired air (and measuring the amount of oxygen in expired air—say 16%) and correcting the volume for a change in temperature, we can calculate the amount of oxygen used by the body to carry out the activity. ISO 8996 (2004) estimates that for a normal mixed diet (RQ = 0.85), 5.68 Wh L^{-1} of O_2 of energy would be produced. If we calculate that 2 L of oxygen were utilized in 5 min, then 24 L of oxygen would be used in 1 h, thus, $24\,L\,h^{-1} \times 5.68\,Wh\,L^{-1} = 136\,W$ of energy is an estimate of metabolic rate. The system of indirect calorimetry is systematic and consistent and has face validity. There are systems of indirect calorimetry other than those described above [full-face mask, breath-by-breath analysis, and other systems where the change in gas content of breathing into flowing air are measured (NIOSH, 2016)]. None, however, are particularly accurate, as they are limited by the assumptions made and the calibrations and practical procedures and precautions required. Parsons and Hamley (1989) suggested that overall an accuracy as low as 50% is not uncommon. This may seem high but acknowledges that training and accuracy required are not always achieved. ISO 8996 (2004) suggests an accuracy of ±5%. Parsons and Hamley (1989) and Parsons and Clark (1984) have pointed out that methods of estimating heat strain in people is particularly dependent upon estimates of metabolic heat production and provides a cautionary note for managers of heat stress when one considers that indirect calorimetry is probably the best practical method of estimation available.

Average Metabolic Rate of a Group

The ventilation rate for an enclosure, such as a room, is related to the rate at which external air enters the room and at which internal air is evacuated from the room. If people occupy the room (ignoring other living organisms, such as plants) and ventilation is insufficient, there will be a "build up" of carbon dioxide which can be measured. In fact, this measure is often taken as an indicator of air quality and ventilation efficacy (ISO 28802, 2012). Using the ventilation rate, the number of people in the room, and the rate of increase of carbon dioxide, we can calculate the amount of oxygen used and the equivalent of burning food to estimate average metabolic rate per person, using a calculation similar to that used in the method of indirect calorimetry. No mouth pieces or other personal equipment is necessary, and hence, there will be no interference with the people involved as they carry out their activities. However, effective mixing of air is assumed, no individual variation is derived, and if carbon dioxide is increasing, then it may complicate the calculation even though the buildup of carbon dioxide is representative of the actual conditions, this means that inspired air percentage of oxygen continuously decreases. This method may have uses for specifying acceptable requirements for enclosures, but further work is required to establish validity in estimating metabolic rate. If ventilation rates are sufficient or greater than required, then there will be no buildup of carbon dioxide. It may be possible to compare input oxygen and carbon dioxide rates and levels to output (exhaust) oxygen and carbon dioxide rates and levels for the room and derive an estimate of average metabolic rate.

However, this will require assumptions to be made about the room. In many contexts, CO_2 does build up. For the purposes of measurement, ventilation may be restricted for the duration of the measurement. The method has the great advantage of noninterference and the use of a simple measure. It has not been proven in general application, however.

Estimation of Metabolic Rate from Heart Rate

There is a linear relationship between heart rate and activity level (hence metabolic rate — ISO 8996, 2004; NIOSH, 2016). That is, from heart rates above resting level, and levels significantly influenced by psychological factors, up to maximum heart rates where maximum oxygen uptake occurs (VO^2 max). Heart rate is also related to hormone distribution and thermoregulation but activity level dominates. The actual heart rate of a person is related to individual factors, and in particular, their fitness, often represented by their VO^2 max. So there is no direct relationship between the average heart rate of a group and average metabolic rate or one relationship between heart rate and metabolic rate for all people. We, therefore, need to determine this relationship for individuals. The usefulness of this method is then to be able to estimate metabolic rate for an individual, from their heart rate, which is easier to measure in practical contexts than using indirect calorimetry.

We can estimate VO_2 max using indirect calorimetry where the subject performs to their maximum capacity (involving shouting encouragement to the subject to maximize motivation). A simplified version is to assume a maximum heart rate and measure oxygen uptake and heart rate at lower levels of activity. Equations vary, but maximum heart rate predictions decrease with age in years with examples as follows: 220 – age; 200 – age; 180 – age, and 205—0.62 × age. By selecting three or so exercises covering say 20%, 40%, and 60% of VO_2 max [resting heart rate + % (maximum heart rate – resting heart rate)] and measuring metabolic rate in a laboratory using indirect calorimetry, an equation can be drawn for an individual. The relationship is for the increase in heart rate and metabolic rate above resting levels. So the measured heart rate minus resting heart rate; divided by the metabolic rate minus resting metabolic rate; is a constant which is determined for each individual. Metabolic rate is then equal to resting metabolic rate + k × (heart rate – resting heart rate), where k is the constant derived as described previously.

This method is particularly useful where it is not practicable to measure metabolic rate using indirect calorimetry. I was asked to investigate heat stress and strain in London underground railway tunnels during the construction of the Jubilee line extension as it ran adjacent to the river Thames (O'Brien et al., 1997). The activity for workers included using hand-held diggers, and it was not possible to use a Douglas bag or other system. The solution was to calibrate the workers in the laboratory as described previously and measure the heart rate in the tunnels using a simple chest strap transducer and "wrist watch" receiver system and to predict metabolic rate from their individual "calibration" equation. The system worked well in normal tunnels. However, some parts of the tunnel used compressed air to keep out the river Thames providing uncertainty to the validity of the equation for those conditions.

Estimation of Metabolic Rate Using Databases and Tables

Most people do not have the skill or access to the equipment to measure metabolic rate. By far, the most common method of estimating metabolic rate for use in heat stress assessment is to refer to tables or databases of previously measured values (usually from values measured using indirect calorimetry). The values are provided in Durnin and Passmore (1967), NIOSH, (2016), ISO 8996 (2004), and Ainsworth et al. (2011).

ISO 8996 (2004) Ergonomics of the Thermal Environment—Determination of Metabolic Rate

ISO 8996 (2004) presents standardized methods and data for estimating metabolic rate which can be used in the assessment of heat stress. Four levels are presented (with increasing accuracy and complexity and requiring a greater level of expertise) according to the risk assessment strategy presented by ISO 15265 (2004). Screening methods include the use of simple tables of data for general level of activity and occupation types. Observation methods include tables of activities described in more detail. An analysis method uses heart rate measurement and the highest "expert" level describes the method of indirect calorimetry as well as a doubly labeled water method and direct calorimetry. The doubly labeled water method allows the estimation of average metabolic rate over long periods (weeks) and is not normally used in heat stress assessment. Individual variability, work speed, skill, equipment, gender, body size, and cultural differences are all identified as possible significant factors. When analyzing a work shift, a time-weighted average of metabolic rates for each activity carried out is recommended as an estimate of overall average metabolic rate, when a range of tasks is performed.

It should be noted that as the body temperature rises, so does the metabolic rate due to the increase in chemical reactions (the Q_{10} effect). This is generally regarded as small and not usually considered in heat stress assessment methods.

The estimate of metabolic rate is highly influential on the outcome of heat stress assessment methods. However, the accuracy with which this can be estimated is probably low. It is debatable whether in practice a standardized value for a general description of activity may be the best that can be achieved. Table 3.1 provides values from ISO 8996 (2004). Whether a conservative estimate is made (confident that the actual metabolic rate is likely to be lower than the estimate of metabolic rate) or a riskier strategy is taken (for some individuals, metabolic rate is likely to be higher than estimated) is a matter for the management of heat stress. This will greatly improve with experience in the context of the assessment.

Metabolic Rate in Load Carriage

There have been studies for particular industries and activities that have produced regression equations to predict metabolic rate values for (usually military) load carriage. These will be for specific contexts and include those of Givoni and Goldman (1971), Pandolf et al. (1977), and Legg and Patemen

TABLE 3.1 Metabolic rate values for categories of activity (watts)

ACTIVITY LEVEL	*METABOLIC RATE (W)*	*RANGE (W)*	*TYPES OF ACTIVITY*
Resting	115	100–125	Sitting or standing at ease. Relaxed at rest.
Low	180	125–235	Light manual work. Hand and arm work. Bench work. Slow walking level surface. Arm and leg work (driving).
Moderate	300	235–360	Heavy bench work. Hammering. Heavy driving. Paced level walking. Picking fruit, hoeing, weeding. Pneumatic hammer. Pushing wheelbarrow. Plastering.
High	415	360–465	Intense arm and trunk work. Carrying heavy material. Shoveling. Sledgehammer work. Sawing. Pushing loads. Fast level walking.
Very high	520	>465	Very intense activity. Running or walking very quickly. Intense shoveling or digging; climbing stairs, ramp, or ladder.

Source: Modified from ISO 8996 (2004).

(1984). For load work with the arms, Morrissey and Liou (1984) and Randle (1987) provide equations. The equations are useful in the context for which they were derived and include variables, such as walking speed and gradient, terrain, box size, body weight, and load weight.

Subjective Methods

An approximation to metabolic rate can be achieved using subjective ratings of how hard people think that they are working. The rating of perceived exertion (Borg, 1982) is the most commonly used scale where ratings are given from "6" as less than "very very light" to "20" which is just above "very, very hard" (9 is very light; 15 is hard, and so on). The values when multiplied by 10 provide estimates of heart rate, and attempts have been made to link metabolic rate values to ratings.

Clothing and Metabolic Rate

For a person wearing protective clothing that is heavier and more cumbersome than normal light work clothing, there will be a metabolic cost. Goldman (2006) suggested, as a rule of thumb that, for an active person, each added layer of clothing will increase metabolic cost by 4%, in addition to its added weight. Teitelbaum and Goldman (1972) and Hanson and Graveling (1999) have measured the values. Dorman and Havenith (2009) provided a comprehensive study over a range of clothing. They found increases of up to 20% over a control suit in both metabolic rate and ratings of perceived exertion. Parsons (2014) provides data from British standard, BS 7963 (2000), which implies an interaction between an increase in metabolic rate and activity. An expert panel suggested no increase for resting in light clothing with safety boots to 15 W m^{-2} for resting in heavy clothing up to an additional 155 W m^{-2} for heavy activity in heavy clothing.

DIFFERENT POPULATIONS AND VULNERABLE PEOPLE INCLUDING PEOPLE WITH DISABILITIES

The metabolic rate for different populations across the world is related to body size. If units are used to normalize for body size (e.g., 58 W m^{-2} for sitting at rest) then for a population where people are relatively small (say 1.5 m^2 average surface area), this will indicate a metabolic rate of $58 \times 1.5 = 87$ W. For a population with 2.0 m^2 average surface area, the metabolic rate will be 116 W. A similar calculation is made for other activity levels. This is an approximation and does not take account of any differences in metabolic heat production that may exist between populations other than due to size (different surface areas).

A similar correction may be made for people with disabilities. However, individual considerations will be necessary. Activities will be altered due to particular conditions and capabilities, and new activities, such as operation of a wheelchair, will become important. There is a shortage of data on metabolic rates for different populations, which provides a limitation for the assessment of heat stress, particularly when related to preserving public health and especially for vulnerable people.

Human Heat Stress and Clothing

4

CLOTHING RESTRICTS HEAT TRANSFER

If a naked person stands in still air, there will be a temperature gradient between the skin and the air that will form an "air layer" and provide insulation to the body. (This is best experienced when an analogous layer forms around a person in cold or hot water.) If the air is moved (by wind or human movement), then the skin temperature will approach the air temperature. Clothing will provide insulation (thermal resistance) between the skin and the environment and reduce or prevent heat transfer. It will greatly influence the strain on people caused by heat stress and if its effects are not recognized can (easily) cause heat illness and death.

As well as air between skin and clothing, an air layer will also form on the surface of the clothing. As well as resistance to dry heat transfer (conduction, convection, and radiation), clothing will provide a resistance to evaporation of sweat (depending upon the vapor permeability properties of the clothing). For any garment or clothing ensemble worn, there will be a dynamic process which is not fully understood but will involve thermal insulation, vapor permeability, wicking, pooling and dripping of sweat, ventilation and pumping, air penetration, compression due to wind, movement and support points, condensation, the transmission, absorption and reflection of radiation, and more. In addition, the body will be divided into those areas covered by clothing and those uncovered (bare skin), and there will often be an uneven distribution of clothing in thickness and in area.

THE THERMAL PROPERTIES OF CLOTHING

Conduction involves heat transfer through a material from high to low temperature. The units of conduction are W °C^{-1} so the units of thermal resistance (insulation) are °C W^{-1} and for each square meter of clothing over the body surface m^2 °C W^{-1}. For convenience (and originally to help military generals understand), we use the term Clo (Gagge et al., 1941) where 1 Clo = 0.155 m^2 °C W^{-1} and is the insulation of a typical male business suit. This should not be confused with a Tog (Pierce and Rees, 1946) where 1 Tog = 0.1 m^2 °C W^{-1} and is a measure of the insulation of material rather than of clothing. The m^2 term for the Tog refers to a square meter of material [usually measured on a calibrated heated flat plate (see also ASTM, 2005a,b)]. The m^2 term for the Clo unit indicates that it is the insulation for a square meter of the human body. This is usually measured by putting the clothing on a heated manikin. The Clo value gives the insulation that a garment or clothing ensemble gives to the whole body whether it is completely covered by the clothing or not. A simple example is that of a necktie. One square meter of necktie material may provide 0.8 Tog of insulation (0.08 m^2 °C W^{-1}). An actual necktie measured on an otherwise nude thermal manikin would give 0.1 Clo of insulation (0.0155 m^2 °C W^{-1}). Thus, for clothing, we measure thermal insulation in Clo. Typical values for the thermal insulation of garments and clothing ensembles are provided in Table 4.1.

TABLE 4.1 Thermal properties of garments and clothing ensembles

Effective Insulation of Garments (for Use in Summation)	Iclu (Clo)
Underwear	0.10 (0.03–0.22)
Trousers	0.20 (0.06–0.28)
Dresses/skirt	0.20 (0.15–0.4)
Boiler suit	0.55 (0.55–0.9)
Sweater	0.28 (0.12–0.35)
Jackets	0.30 (0.25–0.35)
High insulative, fiber-pelt trousers + jacket	0.75
Outdoor coat	0.60
Parka	0.70
Socks	0.05 (0.02–0.10)
Nylon stockings	0.03

(Continued)

TABLE 4.1 (*Continued*) Thermal properties of garments and clothing ensembles

Shoes	0.03 (0.02–0.10)	
Gloves	0.05	
Intrinsic Insulation of Work Clothing Ensembles	**I_{cl} (Clo)**	
Underpants, boiler suit, socks, shoes	0.70	
Underpants, shirt, trousers, jacket, socks, shoes	0.85	
Underwear with short sleeves and legs, shirt, trousers, jacket, thermojacket, socks, shoes	1.25	
Underwear with short sleeves and legs, shirt, trousers, jacket, thermojacket and trousers, socks, shoes	1.55	
Underwear with long sleeves and legs, thermojacket + trousers, outer thermojacket + trousers, socks, shoes	2.20	
Vapor Permeation (Intrinsic Evaporative Resistance of Clothing	**Re,cl (kPa m² W⁻¹)**	
Shorts and shirt	0.010	
Business suit	0.030	
Jeans and shirt	0.020	
Overalls and shirt, work shirt, and trousers	0.025	
Tyvek coverall	0.034	
Gore-tex 2-piece suit	0.033	
PVC suits (polyester, vinyl), neoprene	0.010 (0.094–0.115)	
Ventilation of Clothing	**L min⁻¹**	
Foul weather suit	Slow walking still air/10 mph wind	45.4/108.7
Work pants, polo shirt, and sweater	Standing/ walking 2.5 mph	18.6/127.3

Source: Adapted from Olesen and Dukes-Dubos (1988), ISO 9920 (2009), McCullough et al. (1989), Bouskill (1999), Havenith et al. (1990).

It should be noted that a typical value for the thermal insulation of an air layer around a naked person (I_a) is about 0.7 Clo for still air, decreasing exponentially towards zero as air velocity rises above 5 ms^{-1}, especially if turbulent. At high air velocities, the air is replaced so rapidly that skin temperatures approach the air temperature. The thermal insulation of clothing without including the insulation of the air layer is termed intrinsic clothing insulation (I_{cl}) and with the insulation of the air layer included, the total clothing insulation (I_t). Thus, $I_t = I_{cl} + I_a/f_{cl}$, where f_{cl} is the ratio of the body surface area when clothed to the nude body surface area and takes account of the increase of surface area available for heat exchange when clothed. McCullough et al. (1985, 1989) provide an estimate as $f_{cl} = 1 + 0.31\ I_{cl}$, where I_{cl} is in Clo units.

Table 4.1 presents the thermal properties of garments and clothing ensembles in terms of the intrinsic thermal insulation and evaporative resistance values and data on the ventilation properties of clothing. The properties were measured on thermal manikins and also human subjects for ventilation values. To obtain the intrinsic thermal insulation values for a clothing ensemble, use the table provided for ensembles or select the appropriate garment values and sum the values. Still air has values of 0.7 Clo and 0.014 kPa m^2 W^{-1}. For the effects of ventilation in the heat, reference can be made to a psychrometric chart (Parsons, 2014).

VAPOR PERMEABILITY OF CLOTHING

The transfer of water vapor through clothing is termed vapor permeability. It is important to heat stress assessment as it determines the rate at which sweat can evaporate from the skin and cool the body. The heat transfer by evaporation (vapor transfer) can be regarded as driven by a difference in vapor pressure across the clothing so its units are W kPa^{-1}. Evaporative resistance of clothing therefore has units kPa W^{-1} and for each square meter of body surface area, kPa m^2 W^{-1}. The evaporative resistance of the still air layer around the body (Re,a) is approximately 0.014 kPa m^2 W^{-1} and for a man's business suit, intrinsic evaporative resistance (Re,cl) is about 0.033 kPa m^2 W^{-1}. The system has parallels with the system of units for thermal insulation. Total evaporative resistance (Re,t) is Re,cl + Re,a/f_{cl}. Just as the resistance of the dry air layer around the body is 1/hc, the evaporative resistance of the air layer is 1/he where he/hc is a constant—the Lewis relation (see Chapter 2). There are many indices that describe the vapor permeation properties of clothing, which creates confusion. There is a requirement to clarify the relationships and present vapor permeation in a manner analogous to that for describing thermal insulation as well as the need to clarify the role of radiation in the specification of clothing properties. For a full description see Parsons (2014).

CLOTHING VENTILATION

Gaps, flaps, and openings in clothing as well as air permeability, can alleviate heat stress by directly transferring hot moist air from the skin to the environment. Tracer gas techniques are often used to measure clothing ventilation values. Measurements are represented by Crockford and Rosenblum (1974), Birnbaum and Crockford (1978), Lotens and Havenith (1988); and Bouskill et al. (2002) (see Table 4.1), which will depend upon the type and level of activity (pumping effects). ISO 7933 (2004) provides an analytical assessment of heat stress and notes that activity and ventilation modify the insulation characteristics of the clothing and the adjacent air layer. The standard provides methods (correction factors) in terms of multiple regression equations, involving walking speed and the relative air velocity between the person and the environment.

MEASUREMENT OF THE DRY THERMAL INSULATION OF CLOTHING

A thermoregulatory response to heat stress is vasodilation, which increases skin temperature to make it more uniform across the body and towards internal body temperature. A thermal manikin, maintained at a skin temperature of 34°C (or maybe higher), therefore, represents the human body in shape, size, and skin temperature profile. The power required to maintain a constant skin temperature, when the skin is at a higher temperature than air temperature, in a controlled environment, such as a climatic chamber, in this balanced steady-state condition, is therefore equal to the rate of heat lost to the environment. If air and wall temperatures are constant (e.g., 25°C) and air is still with say 50% relative humidity, we can calculate the thermal resistance of the air layer around the nude manikin. That is the difference between the manikin skin temperature and air temperature divided by the power required to maintain the manikin at the constant skin temperature. In the current example, if it takes 45 W to power the manikin, the insulation provided by the air is $I_a = (34 - 25)/45 = 0.2°C\,W^{-1}$, and if the manikin has a skin surface area of 2 m^2, then the resistance I_a is $0.1\,m^2\,°C\,W^{-1} = 0.67$ Clo. If we maintain identical environmental conditions but dress the manikin in clothing of interest, we will find that after a steady state is reached (a number of hours), it requires less power to maintain the manikin at 34°C skin temperature because clothing has reduced the rate of heat loss (increased the thermal resistance). Suppose it now takes 18 W to maintain the manikin at a constant skin temperature of 34°C. The total insulation is $I_t = (34 - 25)/18 = 0.5$°C W^{-1} or 0.25 m^2°C W^{-1}. That is 1.6 Clo, the total insulation (I_t) of the clothing and the surrounding air. The intrinsic insulation of the clothing is therefore $I_t - I_a/f_{cl}$. If we estimate f_{cl} as 1.5, $I_{cl} = 1.6 - 0.67/1.5 = 1.15$ Clo or 0.179 m^2°C W^{-1}, that is, the insulation of the clothing without influence from the external environment.

MEASUREMENT OF CLOTHING VAPOR PERMEATION, VENTILATION, AND PUMPING

Vapor permeation of clothing can be measured using a sweating manikin. Ventilation and pumping effects can be measured using moving manikins. Simulated sweating using a totally saturated undergarment or by using engineered sweating valves can provide a sweating skin beneath clothing. Calculations involve the power required to maintain a constant skin temperature, and vapor pressure calculations lead to estimates of the total and intrinsic evaporative resistance as well as that of the air layer. To estimate the effects of human movement (e.g., pumping), walking manikins are often used. The technical difficulties in creating sweating manikins as well as manikins that move in a way similar to humans to provide valid simulations are significant, and great expertise and experience is required. Effective standardization that provides for the construction and operation of heated thermal manikins, particularly those that move and sweat, is yet to be achieved. Results can be useful; however, they still depend upon the specific laboratory and the manikin used.

MEASUREMENT AND ASSESSMENT OF THE THERMAL PROPERTIES OF CLOTHING USING HUMAN SUBJECTS

By using a human subject as a manikin, we can investigate the properties of clothing. If environmental conditions are controlled, we can set up a heat balance equation, and by accounting for all avenues of heat loss and gain, we can derive the heat transfer properties of clothing (Kenney et al., 1993). There are particular heat stress conditions where reliable assumptions apply and allow some of the avenues of heat transfer to be eliminated. This helps in the calculations. If wall temperatures, and other surfaces, in a climatic chamber are at the same temperature as the body and clothing temperature, then we can assume that net radiation heat transfer is zero. If air velocity is high, then the surface of clothing and exposed skin will tend towards air temperature and negate the effects of radiation. If skin temperature is the same as air temperature, then heat transfer by convection will be negligible. If a person is performing the same activity, then we can assume that the metabolic rate is similar. A useful assumption in experimental work is to be able to say, despite different conditions, when the thermal strain is the same. We can then look at the relative effects of different variables, including clothing. An equivalent point for the thermal strain of a person is when the internal body temperature is at its maximum point before thermoregulation fails to ensure that sufficient heat is lost to maintain thermal balance. This is termed the upper limit of the prescriptive zone, as beyond that, we enter the environmentally driven zone where at increased heat stress, the internal body temperature will rise, hence identifying an inflection point. For otherwise identical conditions, for example, clothing ensembles can be compared by measuring the temperature where the inflection point occurs. Whatever the combination of environmental conditions, metabolic rate, and clothing, there will be heat balance at the critical point and identical thermal strain. The change in critical point with air temperature or air humidity, for example, can include the influence of clothing, and by determining and equating the heat balance equations for the different conditions, parameters that define clothing properties can be estimated (Kenney et al., 1988, 1993; Garzón-Villalba et al., 2018). Such methods require an accurate estimate of metabolic heat production (and other parameters) in the calculations, however, which can undermine the accuracy of calculations of clothing properties, although not the effects of clothing for a given activity.

The identification and specification of clothing properties are important for the selection and procurement of clothing and the prediction of thermal strain. For practical application, however, we do not need to know the thermal properties of clothing. We only need to know whether the clothing "works" or not. That is, meets its performance requirements. This includes thermal insulation and ensuring that enough sweat will be evaporated for a person to remain comfortable or survive. Another functional requirement (other than thermal) may be to ensure that toxins are prevented from entering the body. As well as functional requirements, there are other requirements (Renbourn, 1972). For that, we should note the four "F's of clothing: function, fit, fashion, and feel" (Goldman, 2006; Goldman and Kampmann, 2007). Fashion is of particular importance and greatly influences choices as well as whether people will be willing to wear clothing or not. The thermal performance of clothing is extremely important in heat stress assessment and can be a matter of life and death. Performance tests in climatic chambers through user trials under actual conditions can provide an indication of the consequences of wearing clothing in hot conditions. Dependent variables include internal body temperature, skin temperature, heart rate, and mass loss by sweat production and evaporation. A performance indicator will be how much sweat is trapped in clothing compared with sweat produced and evaporated. Under compensable conditions (the body can achieve heat balance by thermoregulation), equivalent heart rate may indicate equivalent thermal strain and, hence, show how clothing can influence the ability to carry out physical work during heat stress.

To complement physiological measures, it is important to take subjective measures (Parsons, 2014). These can refer to thermal strain as well as to experiences with clothing. Comparison judgments in different conditions or clothing types are particularly valid. The technical specification of clothing is important for selecting clothing and predicting thermal strain in hot conditions. This provides a basis for user preference tests and trials, which will include other contextual factors and allow experience to be gained with clothing, management systems (including storage, cleaning, and so forth), and the people wearing the clothing.

HEAT STRAIN IN PROTECTIVE CLOTHING AND EQUIPMENT

Wearing protective clothing and equipment, usually to protect against an environmental hazard other than heat, is not pleasant and can be dangerous, and it is essential that people gain experience in wearing it. Prediction methods for predicting the heat strain of people wearing protective clothing and equipment (e.g., face masks and respirators) are not sufficiently accurate to provide other than a starting point for assessment and an approximation. User tests and trials are required to verify the suitability of protective clothing for any activity and context. Further discussion is provided in NIOSH (2016).

SMART CLOTHING

Smart clothing includes active cooling (air, water, and so on), memory materials, phase change materials, nanotechnology, and micropumps and more

(NIOSH, 2016). Artificial intelligence and wearable computers will probably play a role in the future. Active cooling using "compressed" air is particularly useful in voluminous suits that provide a sealed microclimate around the person. They require an air conditioning system nearby, however, and restrict human movement. Active cooling using water or other fluid, often in underwear containing a system of tubes (sometimes with flat surfaces), provides a cooling system. A controlling system is required, but they allow some activity and are particularly useful for people operating vehicles or other workstations. Memory materials change shape with temperature, but when they return to their original temperature, they return to their original shape. The materials under clothing, set to have a shape with a large volume filled with air when above comfortable temperatures, and set to be flat for comfortable temperatures and below, will allow evaporation of sweat and heat exchange when it is required under heat stress. Phase change materials have microcapsules that change state from solid to liquid, or liquid to solid at an appropriate temperature, extracting latent heat from the body when it is too hot and returning it when too cold. Microengineering and material production techniques allow clothing to contain phase change capsules as well as other micromachines for pumping air. Whether such systems have sufficient capacity to maintain the body in thermal balance under heat stress is debatable. An important point concerning all of the "smart" clothing initiatives is that they must complement human thermoregulation. Under heat stress, the body survives by evaporating sweat. If smart clothing restricts this, then any effect must be to lose at least the heat that would have been otherwise lost by sweating, before it can be considered effective (Maley et al., 2017). Another consideration is that, if areas of the body are cooled by smart clothing, then we must consider the signals sent to the brain. A false sense of cooling may be portrayed that restricts necessary thermoregulatory action. For a fuller discussion of smart clothing see NIOSH (2016).

CLOTHING AND BEHAVIORAL THERMOREGULATION

For the prediction of heat stress, it is important to know which behaviors people have available to them. Opening a jacket, for example, for a hot sweating person, can be particularly effective and significantly changes the dynamics of heat transfer. It undermines any heat balance equation or model that does not take that into account. Probably, the most common and effective behavioral action when a person is too hot is to take off clothing. This will depend upon the adaptive opportunity to do so. The optimum, level of, clothing including clothing thermal properties (vapor permeation, ventilation, and design), for any hot environment can be calculated from the heat balance equation.

ISO 9920 (2007) (CORRECTED 2008): ERGONOMICS OF THE THERMAL ENVIRONMENT—ESTIMATION OF THERMAL INSULATION AND WATER VAPOR RESISTANCE OF A CLOTHING ENSEMBLE

A comprehensive database of the thermal properties of clothing, for garments, ensembles, and textiles, as well as a description of parameters and measurement methods, is provided in international standard ISO 9920 (2009). It applies to steady-state conditions and also considers body movement and air penetration. It does not include absorption of water, buffering or tactile comfort, the influence of rain or snow, special protective clothing, such as water cooled or ventilated suits, nor for separate insulation on different parts of the body.

It provides information in appendices on the effects of reflective clothing, posture, seats, pressure, and washing. It also describes methods for determining the thermal properties of clothing using thermal manikins and human subjects.

Different methods of representing clothing properties are compared and presented, of particular note is the i_m value which is the vapor permeability index often used to specify military as well as other clothing (Woodcock, 1962; Parsons, 2014). Further information and databases of the properties of clothing can be found in ASHRAE (1997) and Al-Ajmia et al. (2008) who measured the thermal properties of Arabic clothing.

Measurement of Heat Stress

5

THE SIX PARAMETERS

Human heat stress is the combined interaction of air temperature, radiant temperature, air velocity, humidity, activity level, and clothing on a person. To assess heat stress and predict heat strain, we need to measure or estimate values of all of those six factors. All of the factors are variables which vary in space and continuously change with time; however, they are often defined as single parameters, which are representative of the changing environment and are therefore termed the six basic parameters.

AIR TEMPERATURE (T_A)

In the context of human thermal environments, air temperature can defined as "the temperature of the air surrounding the body that drives heat transfer between the body and the air." It should not be measured too close to the body, as the air temperature will be influenced by body temperature, and not too far away, as that may not be representative of air that is driving heat transfer and affecting the person.

Air temperature is measured using calibrated thermistors or thermocouples. Mercury in glass thermometers are not recommended as breakage will release toxic mercury. Sensors can be affected by both air temperature and radiation. Rapid movement of air across the sensor, such as when whirling a hygrometer, use of fans or in ventilation systems, and external on moving vehicles or outdoor wind, minimizes the contribution of radiation. Shielding the sensors with silvered open-ended cylinders reduces the contribution of radiation, but they must allow a free flow of air across the sensor and must not heat up, causing a reradiation effect.

ISO 7726 (1998) provides a specification of the required accuracy for instruments for measuring air temperature related to heat stress (see also NIOSH, 2016). Air temperature sensors for measuring heat stress should have a measuring range up to 120°C ± 0.5°C guaranteed at least for a deviation of $|T_A - T_R| = 20$°C. Response time should be as short as possible, and a mean value over 1 min is desirable. The sensor should be effectively protected from radiation (ISO 7726, 1998).

MEAN RADIANT TEMPERATURE (T_R)

Mean radiant temperature can be defined for a point or for an object. At any point in a radiant field it is defined as the temperature of a uniform enclosure in which a small black sphere would have the same radiant exchange as it does with the real environment. It can be measured using a black sphere (Vernon, 1932) with a temperature sensor at its center (in fact the original spheres were made from ball cocks used in English toilets). This is called globe temperature and in effect integrates the effects of radiation from all directions (three dimensional) into a single number. The sphere is black to absorb and emit all wavelengths of radiation.

The globe will achieve steady-state temperature when convective heat loss or gain is equal to (balanced by) radiant heat gain or loss. If there is no net radiation gain or loss, globe temperature will stabilize at air temperature. When there is a radiation gain or loss, it is important to correct for air temperature and air velocity as they will also influence globe temperature (ISO 7726, 1998; Parsons, 2014; NIOSH, 2016). The correction for the effects of air temperature and air velocity will depend upon the diameter of the globe (often 150 mm, but smaller less easily corrected globes are often used for convenience). The smaller the globe, the greater the influence of air temperature and air velocity on globe temperature. Any inaccuracies in measurement of air temperature and air velocity will therefore be carried into the correction of globe temperature to obtain mean radiant temperature at a point in space.

The globes are usually made out of copper for a relatively quick response, but the material type will affect only the response time (typically over 15 min for a 150 mm diameter black globe, even for metal) not the equilibrium value, although the value obtained will be affected by any changes during the measurement time. A smaller diameter globe will achieve steady state more quickly than a larger globe.

The mean radiant temperature for an object, such as the human body, will depend upon its shape (and hence body size and posture) which will influence the areas projected towards radiation surfaces. The integration of projected areas (A_p) over all directions provides the area available for radiant exchange between the surrounding environment and the person (A_r). ISO 7933 (2004) uses A_r values divided by the total surface area of the body (A_r/A_D) for standing (0.77), sitting (0.72), and crouching (0.67). Underwood and Ward (1966) used photographic techniques to provide values for a wide range of angles of solar radiation.

ISO 7726 (1998) provides a specification for sensors for measuring mean radiant temperature. For heat stress assessment, sensors should measure up to 150°C ± 0.5°C. Response time should be as short as possible, and a mean value over 1 min is desirable. For globe temperatures, a steady-state value is required. Response time will be greater than 15 min and accuracy may be low for small globes, due to the correction of globe temperature for air temperature and air velocity.

When comparing the direction of radiation, plane radiant temperature (T_{PR}) is often used. That is "the uniform temperature of an enclosure where the radiance on one side of a small plane element is the same as in the non-uniform actual environment." This measures the radiation for the projected area (A_p) over one direction (up/down, left/right, fore/aft). For example, for a standing man, the up/down projected area (A_p/A_r) is low (0.08). Radiant asymmetry is related to discomfort and thermal stress. It is the difference between opposite plane radiant temperatures (e.g., front–back). When considering heat stress in industry with a steel furnace or agricultural work in the sun, most of the radiation will come from one direction, and plane radiant temperature will supplement information provided by mean radiant temperature.

SOLAR RADIATION

The principles of radiation assessment apply in general, indoors and outdoors; however, the sun provides a special case of high intensity, short wavelength (mostly 300–2,000 nm), and specific direction, moving slowly across the sky varying in altitude and azimuth. Sante and Gonzalez (1988) identify three solar radiation terms: direct, diffuse (scattered sky), and reflected (solar from the ground and surfaces). Solar radiation can be measured in terms of its direction, intensity, and spectral content using pyranometers, filter systems, and bands that cast shadows to eliminate direct radiation and measure diffuse radiation or vice versa.

AIR VELOCITY (v)

Air velocity is the speed and direction of air moving across a person. Often an estimate of average air speed is taken. Convective and evaporative heat transfer to and from the body are sensitive to air velocity, and measurement is required to less than $0.1\,ms^{-1}$. Specifications for measuring instruments (anemometers) are given in ISO 7726 (1998) and NIOSH (2016).

Air velocity sensors (anemometers) for measuring heat stress should have a measuring range up to 0.15 ± 0.1 to $10 \pm 0.6\,ms^{-1}$. The levels should be guaranteed whatever the direction of flow. Response time should be as short as possible to measure variations in velocity. An indication of the mean value over 3 min is desirable and, if a time-varying signal is recorded, an estimate of standard deviation, which will allow a calculation of turbulence intensity (ISO 7726, 1998).

HUMIDITY (ϕ)

Humidity is the concentration of water vapor in the air and is often expressed as partial vapor pressure (P_a). Relative humidity (ϕ, or sometimes rh) is the partial vapor pressure in the air divided by the saturated vapor pressure at air temperature (P_a/P_{sa}). The dew point (T_{DP}) is the temperature at which the air would become saturated, below which water would condense (clouding up cold windows, face masks, or goggles) and it would start to rain.

Partial vapor pressure can be measured directly using calibrated electronic (capacitance) sensors or can be derived from a psychrometric chart using aspirated (sensors in high air velocity) wet- (T_{WB}) and dry-bulb (T_{DB}) temperatures, from a whirling hygrometer, for example. The hygrometer contains two thermometers: one with a soaked (continuously from a distilled water reservoir) wet wick covering its sensor and the other with an open dry sensor. By whirling (rotating longitudinally) the thermometers at high speed in a frame with a rotating handle, the effects of radiation become negligible so dry bulb temperature (T_{DB}) becomes air temperature (T_A) and the wet wick evaporates water and cools the sensor to a temperature that is related to the humidity in the air. A psychrometric chart (Ellis et al., 1972) provides partial vapor pressure, relative humidity, and dew point (see Figure 5.1). Dew point can also be measured optically by cooling a silvered surface until dew forms (hence deflecting a reflected light beam). ISO 7726 (1998) and NIOSH (2016)

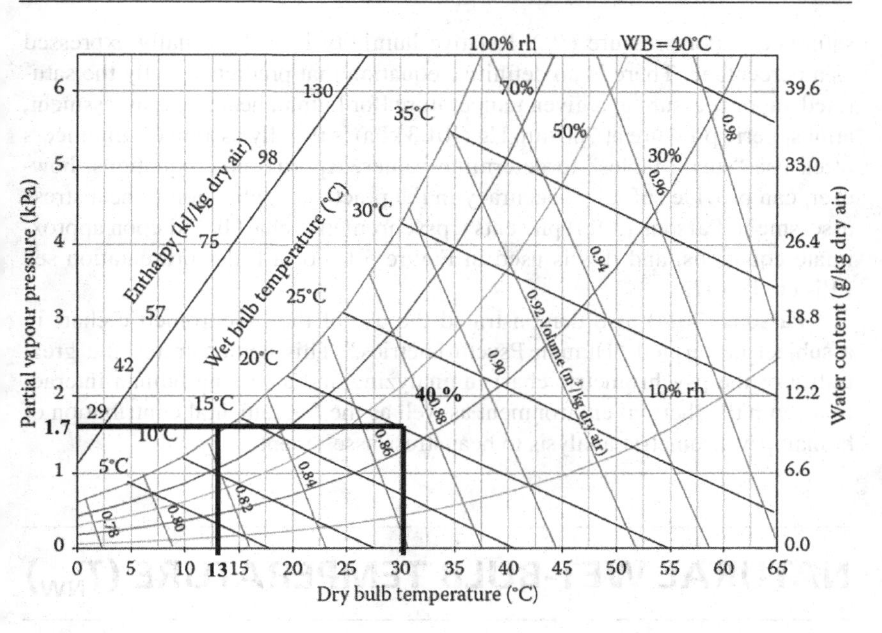

FIGURE 5.1 Psychrometric chart showing how to determine relative humidity, partial vapor pressure and dew point from dry bulb temperature (air temperature), and aspirated (psychrometric) wet-bulb temperature. Further information on human psychrometrics is provided in Parsons (2014). A dry bulb temperature (=air temperature) of 30°C and psychrometric (aspirated) wet-bulb temperature of 20°C, gives 40% relative humidity; 13°C dew point; and 1.7 kPa partial vapor pressure.

provide a specification of instruments for measuring humidity in the assessment of heat stress.

Absolute humidity is measured as the partial vapor pressure of water vapor. Sensors for measuring heat stress should have a measuring range up to 6 ± 0.15 kPa and as short a response time as possible (ISO 7726, 1998).

THE PSYCHROMETRIC CHART

For a given (atmospheric) pressure, the amount of water the air can contain as water vapor increases with its temperature. The water vapor in the air exerts a pressure called the partial vapor pressure (P_a), and when the amount of water vapor is the maximum, the air can hold at that temperature, it is called the saturated vapor pressure (P_{sa}). Relative humidity is P_a/P_{sa} usually expressed as a percentage. There is no definitive equation that predicts exactly the saturated vapor pressure at a given temperature [for human heat stress assessment, atmospheric pressure at 760 mm Hg (101.3 kPa) is usually assumed]. Engineers often use "steam tables" to determine values. Approximate equations, however, can provide sufficient accuracy in the practical application of heat stress assessment. Parsons (2014) presents a psychrometric chart based upon approximate equations, and this is used in Figure 5.1. For a fuller presentation see Ellis et al. (1972).

Parsons (2014) also demonstrated the use of the psychrometric chart in a subject he termed "Human Psychrometrics." This demonstrated the great utility of the psychrometric chart in analyzing and assessing human interactions with the thermal environment as well as the fundamental contribution of human psychrometric analysis to heat stress assessment.

NATURAL WET-BULB TEMPERATURE (T_{NW})

All objects, including temperature sensors, obey the laws of thermodynamics, so relatively hot objects cool down, cool objects heat up, and wet objects evaporate water until an equilibrium position is reached. A temperature sensor covered in a wet (saturated) wick (constantly supplied with a reservoir of distilled water) will reach equilibrium where heat transfer by convection, radiation, and evaporation will add to zero. As this will depend upon the air temperature, radiant temperature, air velocity, and humidity, a natural wet-bulb sensor is affected by the same environmental variables as those of a sweating person. The natural wet-bulb temperature (T_{NW}) is not artificially ventilated [and should not be confused with the aspirated (psychrometric) wet-bulb temperature (T_{WB})], and there is no metabolic rate or any other heat production to consider. The specification for a natural wet-bulb thermometer is presented in Chapter 7. Although natural wet-bulb temperature does not provide values for any of the six basic parameters, it is used as an integral part of the heat stress assessment method involving the wet-bulb globe temperature index (WBGT—ISO 7243, 2017). Parsons (2006) describes a method of adding a simple cylindrical cloth of the clothing worn by people exposed to heat stress to the natural wet bulb, to simulate the clothed sweating person. This provides a simple, probably more valid, and practical way of providing WBGT values that take account of clothing and can be compared with limit values provided in ISO 7243 (2017). A heat balance equation for the calculation of natural wet-bulb temperature from environmental conditions is provided in ISO 7243 (2017).

OPERATIVE TEMPERATURE (T_O)

Operative temperature is the weighted average of air temperature and mean radiant temperature. The weighting factors are the heat transfer coefficients for convection and radiation, respectively [$T_O = (hc \cdot T_A + hr \cdot T_R)/(hc + hr)$]. Approximations to operative temperature can be made by measuring the globe temperature of a black sphere of appropriate size. This misses the point, however, and operative temperature should be considered as a calculated temperature that represents the combined effects of air temperature and radiant temperature or convection and radiation. That is, the temperature that represents the driving force for dry heat transfer. As this is a single number that represents the effects of two relevant variables, it is a thermal index.

THE HEAT STRESS INDEX

A heat stress index is a single number that represents the combined effects of factors that determine heat strain. That is the environmental factors of air temperature, radiant temperature, air velocity, and humidity. Metabolic rate and clothing are sometimes included in the index or the index can be interpreted in the context of metabolic rate and clothing as well as other factors, such as the degree of acclimatization of people exposed to heat. Blazejczyk et al. (2012) provide a comparison of different heat stress indices.

The perfect index will respond to heat in perfect correlation with the heat strain on a person. The greater the heat strain, the higher the value of the heat stress index. The advantage of using an index is that we can use a single number to describe heat strain, rather than four or more parameters with numerous combinations, and use it to describe severity, likely heat strain, and limits to avoid heat casualties. There has been much research into defining the perfect heat stress index and many have been proposed. They have been used across the world, often developed in specific countries and contexts (coal mining, the steel industry, military operations, and so on). For a fuller description see McIntyre (1980), Goldman (1988), Parsons (2014), and NIOSH (2016).

Parsons (2000, 2005) and NIOSH (2016) describe three types of indices. These are direct, empirical, and rational. A direct index uses an instrument that responds to the same combination of factors as do people under heat stress. The temperature of the instrument (or the weighted average of a collection of instruments) exposed to the heat stress is usually taken as the index value [e.g., wet globe temperature (Botsford, 1971); wet and dry temperature (Lind, 1963); WBGT (Yaglou and Minard, 1957)]. Empirical research, measuring heat strain on human subjects at different heat stress levels of the index, is conducted to establish what levels of the index produce what levels of heat strain (internal body temperature, sweat loss, heart rate, and so on). By far, the most accepted direct index used throughout the world is the WBGT index (see Chapter 7). Direct indices are simple, valid but not perfect, easy-to-use measures. Attempts to complicate them with additional corrections are often proposed but start to miss the point.

Empirical heat stress indices are derived by collecting data on human responses to combinations of relevant variables (air temperature, radiant temperature, air velocity, humidity, clothing, and activity) and building up a database of information (often represented in the form of graphs or nomograms). The data are compared with standard conditions to provide equivalent heat strain and used to produce a single number (usually equivalent temperature or sweat rate). For example, the Effective Temperature Index (ET) is the temperature of a standard environment (air temperature = mean radiant temperature; 100% relative humidity; still air) that would give the same thermal strain as in the actual environment. Laboratory studies (investigating "warmth" and later used for heat strain), started with the relative contribution of air temperature and humidity in two types of clothing [Effective Temperature (ET), Houghton and Yagloglou, 1923] and expanded to include air velocity and radiation (Corrected Effective Temperature, Vernon and Warner, 1932). The predicted four-hour sweat rate (P4SR) was developed in London and Singapore (McArdle et al., 1947) and more have been proposed (Parsons, 2014). Empirical indices have the advantage of validity; however, to be perfect they would have to include an infinite number of combinations of variables. An interesting variation is the computer database model first proposed by Parsons and Bishop (1991) which used an actual database of human responses and a "best fit" (between conditions of interest and those in the database) engine to predict heat strain.

Rational heat stress indices are analytical and use the body heat equation to determine requirements for heat balance (Figure 1.1). For heat stress assessment, the index is usually derived from a calculation of how much sweat is required to balance the "equation" (Sw_{req}—see Chapter 8), but required clothing (IREQ) is used for cold stress assessment (ISO 7933, 2004; ISO 11079, 2007). For practical application, required environmental conditions can be determined, such as required air temperature or required air velocity (where fans are the only practical economic option) or a combination. As a heat stress index, the Sw_{req} value is a single number that varies with heat strain.

Partly due to usability and partly due to the increased ease of computer modeling and calculation, the two types of indices used internationally are direct (WBGT) and rational (Sw_{req}). These are described in Chapters 7 and 8.

A development of the thermal index is to use a simulation of the human body under heat stress. Heated, human-shaped manikins provide direct physical models and multiple regression equations can provide empirical models. Rational models use a computer simulation of the human body and its system of thermoregulation. These are described in Chapter 9.

6 Measurement of Heat Strain

PHYSIOLOGICAL INDICATORS OF HEAT STRAIN

The response of the body to heat stress is called heat strain, and as people are homeotherms, any increase in their core temperature will be an indication of heat strain. The term core temperature is a concept related to descriptions of human thermoregulation. It does not exist in practice and should not be used when describing measurements. It is the temperature that the body attempts to defend in homeostasis and can be contrasted with "shell" temperature which varies in an attempt to maintain core temperature. There are areas of the body (internal structures and space, the brain, and vital organs) that can be regarded as "core," and it is the tendency for their temperature to rise, if no action was taken, that defines heat stress. Internal body temperature, skin temperature, heart rate (as a general indicator of strain), and sweat rate (particularly in heat stress) are the four main physiological measures of heat strain.

INTERNAL BODY TEMPERATURE

Internal body temperature is controlled by thermoregulation, so if it is rising above a "set point" equilibrium position, then there is heat gain that can lead to confusion, heat injury, and death. ISO 9886 (2004), Parsons (2014), and NIOSH (2016) describe aural temperature (in the ear canal); tympanic temperature; oral temperature; esophageal temperature; subclavian temperature; intra-abdominal temperature; rectal temperature; urine temperature, and transcutaneous deep body temperature, as areas of the body and mechanisms for measuring internal body temperature.

Although all measurements made in those areas are attempting to measure the internal temperature of the body, they are all different. Each method has its pros and cons in representing the "core" temperature, and there are practical issues and precautions in their measurement (see ISO 9886, 2004). For heat stress, it is important to monitor the internal body temperature for health, and brain temperature is probably most important. Measuring in two ears (thermistors insulated from the external environment) and oral temperature (mouth closed on sensor for at least 4 min) are useful. Rectal temperature can give a reliable measure of the internal temperature of the main body mass and is often used by physiologists, but it is not acceptable to some people. Invasive techniques must be used with care and, as with all investigations involving human subjects, have ethical clearance (ISO 12894, 2001).

There are often practical issues (if detecting auditory signals is important then avoid using aural sensors), and expertise is required in fitting and administering equipment and interpreting values. The use of infrared sensors in the ear to measure the temperature of the ear drum is convenient, but a continuous signal is not achieved so the monitoring capabilities are limited. There is some evidence that while useful in moderate environments for measuring the internal temperature of resting patients, infrared systems are limited when used in extreme environments, and for moving, sweating subjects where the value provided may not be valid and the external electronics may malfunction due to high temperatures. Calibration and pretrial are therefore necessary in heat stress environments. An increasingly used method of monitoring internal body temperature is to use a temperature pill. Although this has the advantage of noninterference once taken, taking (often the night before exposure) and recovering of the pill is sometimes not attractive to people and the signal received will vary throughout the alimentary canal. Often, little advantage is therefore provided over less invasive methods.

SKIN TEMPERATURE

When the body is under heat stress, vasodilation increases skin temperature towards "core" temperature and the skin temperature across the body becomes homogeneous. Mean skin temperature is therefore an indication of heat strain and a rise from a comfortable 33°C to 36°C and above is typical. If mean skin temperature rises above core temperature, then heat loss is difficult and the heat stress will produce unacceptable heat strain.

Mean skin temperature is the average of all skin temperatures over the body. One method of measuring this for a naked person is to use infrared photography and integrate the values across the body. Great care is required, however, as all skin cannot be seen by one "photograph," assumptions must be made about emissivity and careful calibration is required. Another method is to use thermistors to measure temperature at selected sites on the skin. Using a large number of sites across the body and selecting those sites which best predict the mean skin temperature as indicated by the larger number of sites reduces the number of measurements to practical feasibility. As skin temperature is homogeneous when a person is hot, fewer sites are required than if the skin had a wider variation in temperature across the body such as is found in cold conditions.

Methods of estimating mean skin temperature are weighted averages, of temperatures at each site, according to the relative body segment size, mass, or surface area. They vary from 1 point methods (medial thigh) to the ISO 14 point method (ISO 9886, 2004) and more (Parsons, 2014). The Ramanathan (1964) four-point method is often used and is appropriate for assessing heat stress responses. For that method, temperatures are weighted as 0.3 × left upper chest + 0.3 × left front shoulder + 0.2 × right interior thigh + 0.2 × right shin. The ISO four point method uses 0.16 × left hand + 0.28 × (right shin + neck + right scapular) (ISO 9886, 2004).

Skin temperature provides some information regarding heat strain, but as it requires sensors to be placed under clothing, it is inconvenient to measure on active people, sweating skin causes difficulty in maintaining contact, and as the body is hot, the skin is homogeneous in temperature and is a less important indicator of heat strain than other measures. (Skin temperature is however an important indicator of cold stress.)

HEART RATE

Heart rate is measured in beats per minute (bpm). ISO 9886 (2004) suggests that overall heart rate can be considered to be made up of the sum of components due to rest, activity, static exertion, thermal strain, psychological reactions, and other factors such as breathing. Heat stress will therefore add to the overall heart rate when compared with conditions where there is no thermal strain and is usefully monitored to ensure that overall strain does not become unacceptable. Some expertise in interpretation of heart rate values is required. An overall heart rate, under heat stress at rest, of 140 bpm indicates much greater strain than 140 bpm say during activity. Heart rate is a convenient and valid indicator of general strain, and if the relationship can be established between heat strain and heart rate for an individual (including recovery times), heart rate can be an effective indicator of heat strain.

SWEATING—BODY MASS LOSS

When the body is under heat stress, it sweats to lose heat by evaporation. In a clothed person, some of the sweat evaporates off unclothed skin; some transfers through clothing as vapor to the clothing surface and beyond; and some transfers heat due to pumping, dripping, and ventilation.

There are three important measures. Sweat produced, sweat trapped in clothing, and sweat evaporated. These can be measured in a systematic way using scales calibrated to an accuracy of around 1 g. It is important when weighing "subjects" to ensure that any equipment worn (for measuring skin, heart rate, internal body temperature, and so on) is not counted as sweat or body mass. As "subtraction" calculations are made, it is important to subtract out the mass of the equipment. Any tape or sensors weighed before the exposure to heat must be weighed afterward. It is a good idea to recommend a 'bathroom' visit before the experiment begins to avoid additional measurements and calculations. First, the subject is weighed nude (usually minimally clothed in slight underwear to preserve modesty) before the exposure (N_B), then the subject puts on clothing and is weighed again before the exposure to heat (C_B). The subject is then exposed to heat stress according to the experimental or work protocol. Immediately after the heat stress exposure, the subject is weighed clothed (C_A) and then the clothing is quickly removed and the subject is weighed nude (N_A). Note that any drinking during the experiment must be accounted for in terms of mass gain and must be subtracted from the mass measurements taken after the experiment. Measuring mass loss during an exposure to heat may not be useful as some of the sweat will be trapped in clothing, but it will give a measure of the sweat evaporated.

From the above measures, $(N_B - N_A)$ provides the total mass loss from the body (mainly and assumed to be sweat produced). This is an indication of the thermal strain as well as dehydration and water requirements. $(C_B - N_B)$ is the dry weight of clothing. $(C_A - N_A)$ is the wet weight of clothing. $(C_A - N_A) - (C_B - N_B)$ is the sweat trapped in clothing, and $(N_B - N_A) - [(C_A - N_A) - (C_B - N_B)]$ is the sweat evaporated.

The sweat evaporated shows how much heat is lost by evaporation and hence contributes to alleviating heat stress. The ratio of sweat evaporated to total sweat produced (*e*/*p*) is often regarded as a measure of the effectiveness of clothing for the activity under heat stress. If all the sweat produced is evaporated, then clothing is very effective. If clothing is impermeable, little sweat will be evaporated and sweat will pool. Clothing will be ineffective for heat loss although some heat will be lost by condensation and conduction, it will be a relatively small amount.

PERSONAL MONITORING SYSTEMS

Management systems are required to ensure health and safety when people are exposed to heat stress. ISO 15265 (2004) and ISO 12894 (2001) provide risk assessment strategies and screening methods to avoid exposing vulnerable people to heat stress. NIOSH (2016) describe the commercially available personal monitoring systems. People exposed to extreme heat are particularly at risk, and their internal body temperature and heart rate should be monitored. Personal monitoring systems, integrating a number of measures, and often transmitted to a central control, are often used to measure these and other factors. The method of interpretation of responses should be established before exposure and performed by experienced experts. Limits will depend upon context but increase in heart rate due to heat or towards maximum heart rates (e.g., 180—age in years) provide warnings. How quickly someone can be withdrawn from the heat to cool down, especially cease activity and remove clothing, is important. Depending on the activity level, an internal body temperature of 38°C–38.5°C starts to become of concern especially if it continues to rise as heat storage can build up rapidly. Sweat rates are difficult to measure continuously. Dripping can be observed, and sweat rates can be measured over shifts. Color of urine should also be monitored where a dark color can indicate dehydration. Mean skin temperature is cumbersome to measure and not essential but is useful to support interpretation of internal body temperature measures, if available. The dangers of wearing protective clothing and equipment should not be underestimated even in conditions (temperatures) that would not normally be considered to cause heat stress. Personal monitoring systems are viable methods for ensuring the avoidance of heat casualties. Technology has ensured that many variables can be measured, but it must be remembered that expertise is required in producing a monitoring, management, interpretation, and response plan.

PSYCHOLOGICAL MEASURES

Heat stress will affect human mood, comfort, and behavior and can range from depression, aggression, and avoidance (Baron, 1972), to elation and self-imposed increased exposure with dangerous consequences. A well-dressed lady at a dinner party does not want to sweat, but a runner or a soldier on a military exercise may be happy to sweat profusely as evidence of effort and achievement. Interestingly, runners that train at altitude have complained of lack of sweating when in fact sweating has occurred but the low atmospheric pressure has caused rapid evaporation to occur.

Behavior can be measured by observation in a formal way by recording behaviors and environmental conditions. An example is the use of a two-way mirror to study the behavior of school children in hot classrooms (Wyon, 1970; Humphreys, 1972). Behavior can also be studied in retrospect, such as the development of riots in crowds in hot weather. Aggressive acts from the police (security forces or military personnel, exacerbated by uniforms and equipment) stimulate aggressive acts from a crowd, and there is incentive for people in a crowd to run into spaces to avoid heat stress by being squashed together. A case was reported of a man who removed all of his clothing in the United Kingdom heat wave of 1911 while walking from one village to another. For more typical behavior in heat stress, observation of adjusting or removing clothing garments in hot conditions can be measured. As with all behavioral measures, establishing cause and effect is difficult, but there is no question that heat stress can affect behavior and that it can be measured. Irrational behavior will be an indication of confusion and a warning of danger. Those responsible for the safety of people in the heat should develop systems to identify irrational behavior including loss in cognitive ability.

Subjective assessment should allow people to remove themselves from heat stress but not to stay in it. Independent, preferably noninterested, trained personnel should decide whether it is safe to continue with exposure, based upon explicit criteria. There are many subjective scales that can be used to measure the effects of heat stress from discomfort to stickiness to acceptability and tolerance (ISO 28802, 2012). Borg (1998) provides a well-researched scale of exertion. It ranges from 6 to 20, where multiplying the rating by 10 gives a rough estimate of heart rate.

7 Assessment of Heat Stress Using the WBGT Index

THE WET-BULB GLOBE TEMPERATURE INDEX

Of all of the heat stress indices, the wet-bulb globe temperature (WBGT) index is the most widely accepted and used throughout the world (Yaglou and Minard, 1957; Parsons, 2014; NIOSH, 2016; ISO 7243, 2017; ACGIH, 2018). It is a simple direct index based upon the weighted average of temperatures. There are others of similar validity that have been developed (e.g., wet globe temperature, Botsford, 1971) but the WBGT has been standardized and used internationally such that important experience has been gained with its use. That is, in heat stress assessments from military, security, and industrial applications, to leisure, sport, and more. It is used increasingly in the management of sports events where exertional heat stroke is a serious problem (ACSM, 1984).

It all started around the early 1950s in the United States when a commander in the U.S. Navy and a researcher from the Harvard School of Public Health, along with a team of investigators, set up a system for reducing heat casualties in military bases (Yaglou and Minard, 1957). Air temperature and humidity alone had been used as indicators of when it was safe to train in the heat, but there were unacceptable levels of heat casualties. A current index at the time, the corrected effective temperature (CET) is the Effective

Temperature corrected using a globe thermometer reading for radiation, termed Effective Temperature Radiation (ETR) in the study (Houghton and Yagloglou, 1923; Vernon and Warner, 1932) was found too cumbersome for practical application. The solution was the integrated temperature of a wet-bulb thermometer (T_{NW}), a black globe thermometer (T_G), and air temperature (T_A) giving the WBGT. This provided an approximation to the CET (ETR) value and is directly influenced by air temperature, radiant temperature, air velocity, and humidity. That is, the same basic parameters that determine heat stress on people and hence providing face validity. Although originally derived for outdoor activity, the index is now used to assess heat stress both indoors and outdoors.

The simple WBGT measuring instrument was not strictly defined at first but is now standardized, most recently in ISO 7243 (2017) (see Figure 7.1). A weighted average of sensor values provides WBGT $= 0.7 \times T_{NW} + 0.3 \times T_G$. Both sensors are exposed to the air temperature, radiation, air velocity, and the humidity of the environment, but the equation is not applicable when the person and hence sensors are in direct sun. A second equation is used, also applicable indoors and outdoors, where there is direct solar radiation. That is WBGT $= 0.7 \times T_{NW} + 0.2 \times T_G + 0.1 \times T_A$.

There have been a number of variations and interpretations of the WBGT measuring instrument and formulae which are incorrect and should not be used other than for approximation. That is, even when corrections are attempted, to compensate for what is used, to estimate values of what ought to have been used. A common example is the use of a black globe of smaller

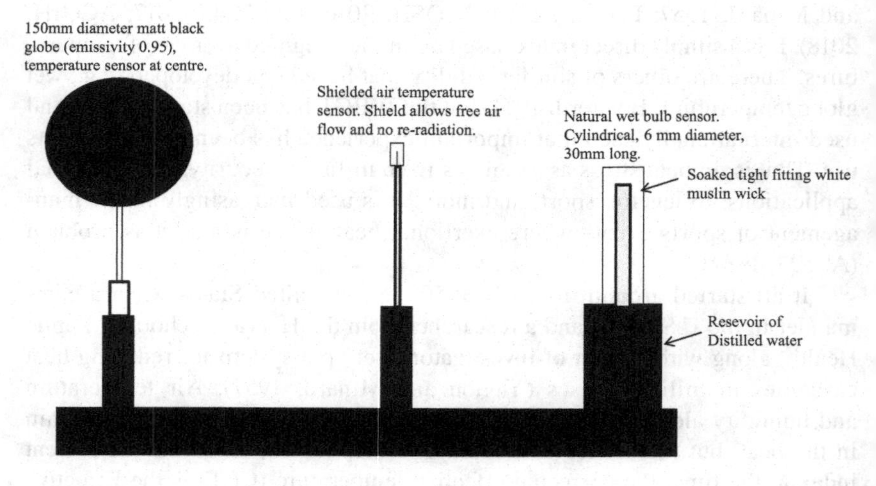

FIGURE 7.1 The WBGT measuring instrument (not to scale).

diameter than 150 mm. The effectiveness of corrections will depend upon environmental conditions. A high air velocity and radiation environment will provide significant error when the instruments used are not up to the standardized specification but may show insignificant error for still air and negligible net radiation.

Globe temperature diameter is a common variation as the specified diameter of 150 mm (6 in.) is often inconvenient to use and transport. Smaller globes (25–125 mm are typical) are greatly affected by air velocity in a way that is difficult to measure accurately and negates one of the original stated advantages of using the WBGT index that air velocity is not required (Yaglou and Minard, 1956).

Yaglou and Minard (1956) report that they used the cloth from the marine-corps trainees' uniform, to cover the black globe thermometer, to simulate the effects of color of clothing (olive drab herringbone twill fabric) on (solar) radiation absorption. An early form of the WBGT equation for use in the sun was $WBGT = 0.7\ T_{NW} + 0.3\ (\alpha\ (T_G - T_A) + T_A)$, where α is the solar absorptivity of clothing. For a black body, or no solar radiation but maybe longer wavelength radiation (indoors), $\alpha = 1.0$ so $WBGT = 0.7\ T_{NW} + 0.3\ T_G$. For green clothing, α is around 0.67 in the sun (actually for olive drab uniform α is given as 0.74) so $WBGT = 0.7\ T_{NW} + 0.2\ T_G + 0.1\ T_A$ for solar radiation as the effects of color of clothing will be significant.

The natural wet-bulb temperature (T_{NW}) has been standardized as the same measure to be used in both formulae, that is, in and out of solar radiation. However, Yaglou and Minard (1956) report the use of psychrometric (aspirated and hence not significantly affected by radiation) wet-bulb temperature (T_{WB}) for the formulation, applicable to conditions with no direct solar radiation, and natural wet-bulb temperature (T_{NW}) where there is solar radiation. International standard ISO 7243 (2017) does not use psychrometric wet-bulb temperature and standardizes the use of natural wet-bulb temperature for all contexts and exposed to whatever long or short wavelength radiation is present. It should be noted that at high air velocities, such as those in outside wind, the natural wet-bulb temperature will tend towards the psychrometric wet-bulb temperature as the wind forces evaporation towards a maximum.

AVOIDING HEAT CASUALTIES IN THE U.S. MARINE CORPS

On 31st May 1956, the Office of Naval Research, Physiology Branch, Washington DC, United States, published a final report on a study, which took place in the early 1950s, entitled, "Prevention of heat casualties at marine corps training centres." By the conclusion of the study, heat casualties had reduced from an average of 126 per 10,000 recruits to 2.5 per 10,000 recruits and it had spawned a new pragmatic heat stress index, the WBGT, which is now used internationally in the prediction, reduction, and prevention of heat casualties. The principal authors of the report were Constantin P. Yaglou from the Harvard School of Public Health and David Minard from the Naval Medical Research Institute, Bethesda, Maryland. They were assisted by Paul Stonestreet, Kenneth Fennell, and Ann Messer with support from H S Belding from the University of Pittsburgh.

The object of the research was to study the conditions under which heat injury occurs in basic and advanced trainees and to develop safe limits for physical exertion in the heat in order to reduce casualties in hot weather. Heat casualties were significant and the use of air temperature and humidity alone to decide on whether training should go ahead had been deemed inadequate. Three Marine Corps schools were investigated. Quantico in Virginia with an average daily temperature of 27.1°C (80.8°F); Camp Le Jeune in North Carolina 28.5°C (83.3°F), and the Marine Corps Recruit Depot, South Carolina 29.7°C (85.4°F). Thousands of troops (aged 17–23 years) were investigated in a range of studies. For example, in one study in Camp Le Jeune in the summer of 1954, the incidence of heat cases among 5,070 unacclimatized new reservists on a 2-week training duty, was compared with that of 1,246 recruit graduates on a 4-week advanced training programme.

The most common heat illness encountered at the three camps was heat exhaustion. This was characterized by dizziness, lack of coordination, profuse sweating, pallor, headache, dyspnea (shortness of breath), and gastrointestinal disturbances. Heat rash was prevalent, but heat cramps and heat stroke were relatively rare. Exercise of relatively long duration and moderate intensity, as opposed to short duration and high intensity, was found to be the main determinant of heat stress (strain) and evaporative sweat loss correlated well with WBGT levels. Heat exhaustion occurred more in recruits from the northern states and at bases further south due to increased heat. The WBGT index was found to be a more effective indicator of heat stress than using air temperature and humidity alone. Previously, there had been unnecessary restrictions to training on cloudy windy days and permitted training on sunny windless days when most heat casualties occurred. Recommendations were made for limit values depending upon the recruit experience and exposure, including a cessation of all physical training at WBGT levels of 31.1°C (88.0°F).

The WBGT index is still used today to regulate exposure to heat of U.S. Marine Corps personnel. A system of flags, flown in Marine Corps installations, is used which indicate increasing risk of heat cramps, heat exhaustion, and heat/sun stroke. For WBGT less than 26.7°C (80°F), a white flag is raised to 29.4°C (85°F), green to 31.1°C (88°F), yellow to 32.2°C (90°F), red and black to above 32.2°C (90°F). Wearing of body armor or NBC (Nuclear, Biological, Chemical) uniform adds approximately ten points (°C) to the measured WBGT index. The black flag condition is determined by commander's decision rather than the WBGT as it is used in exceptional circumstances. The red flag generally indicates cessation of physical work.

ISO 7243 (2017): ERGONOMICS OF THE THERMAL ENVIRONMENT—ASSESSMENT OF HEAT STRESS USING THE WBGT INDEX

International standard ISO 7243 (2017), based upon the WBGT heat stress index, has been adopted across the world, including the 28 countries of the European Union and in legislation in a number of countries (e.g., Sweden and Japan). It presents a screening method for evaluating the heat stress to which a person is exposed. It applies to the assessment of indoor and outdoor environments and to male and female adults who are fit for work. It applies to occupational environments as well as to other types of environment. If the evaluation provided by the use of the standard indicates that there is a risk of heat casualties (WBGT limits exceeded), then reduction in heat stress exposure may be recommended and reference is made to ISO 7933 (2004), which provides an analytical and more comprehensive assessment (see Chapter 8).

The WBGT is defined as "a simple index of the environment that is considered along with metabolic rate to assess the potential for heat stress among those exposed to hot conditions." It combines the measurement of natural wet-bulb temperature (T_{NW}) and black globe temperature (T_G) using the formula $WBGT = 0.7 \times T_{NW} + 0.3 \times T_G$ for conditions without solar load. With solar load $WBGT = 0.7 \times T_{NW} + 0.2 \times T_G + 0.1\ T_A$ where T_A is air temperature.

The black globe sensor is of 150 mm diameter, matte black (emission coefficient 0.95), with a range of 20°C to120°C and accuracy ±0.5°C up to 50°C and ±1.0°C from 50°C to 120°C. The natural wet bulb is cylindrical (6 ± 1 mm diameter and 30 ± 5 mm long) with a measuring range 5°C to 40°C and accuracy ±0.5°C. The support is 6 mm in diameter and a clean white (cotton) wick sleeve covers the whole of the sensor and 20 mm of the support. The air temperature sensor shall be shielded from radiation by a device that does not re-radiate radiation and allows free air movement. The measurement range is 10°C to 60°C with an accuracy of ±1°C.

The assessment method requires measurement to be made over 1 h that represents the heat stress exposure (instruments may have to be used for longer than that time to get a settled reading). If the heat stress varies so that a single value does not represent the heat stress exposure of the person (divides into distinctly different types or categories), then more than one assessment is made. Heat stress assessment should be made at the height of the abdomen (or at the position of most heat stress when there is heterogeneity) and at a position in space representative of the environment to which the people of interest are exposed. Where there is variation in conditions, activity, and clothing throughout the hour of assessment, a time-weighted average value is calculated to obtain final WBGT, clothing, and metabolic rate values.

WBGT limits are provided for people wearing standard light clothing ((boiler suit/cloth coveralls or long sleeve cotton shirt and cotton trousers (pants); $I_{cl} = 0.6$, $i_m = 0.38$ (vapor permeability of cotton)). If the clothing varies from the standard clothing then a clothing adjustment value (CAV) is added to the WBGT value to provide the effective WBGT ($WBGT_{eff}$) which is used in the assessment. The $WBGT_{eff}$ value is compared with a WBGT reference value (exposure limit) which depends upon the metabolic rate and whether the people are acclimatized or not. An acclimatized person is one who has been exposed (presumably without mishap) to the conditions at least as severe as those being assessed for at least the previous full working week. If not, the person shall be considered unacclimatized.

Metabolic rate values are provided in ISO 8996 (2004), and the classification of levels of metabolic rate is also provided in ISO 7243 (2017). Central values for categories are provided in Watts (resting, 115; low, 180; moderate, 300; high, 415; very high, 520). CAVs (Bernard et al., 2008) are in °C WBGT and are 0 for clothing equivalent to the standard clothing and between 2–4 for heavier permeable clothing and 10–12 for vapor barrier clothing. A 1°C WBGT is added for the wearing of a hood of any type.

Limit WBGT reference values (exposure limits)) for $WBGT_{eff}$, below which heat casualties would not be expected, are provided in three forms for both acclimatized and unacclimatized people. The first is by category of metabolic rate. For resting people (115 W), the WBGT reference limits for persons acclimatized/unacclimatized to heat are 33°C/32°C WBGT, respectively. Values are provided for classes 0 (resting) to 4 (very high) as follows. Resting 33/32; low 30/29; moderate 28/26; high 26/23, and very high 25/20.

The second form of WBGT limits is where an estimate of metabolic rate is more accurate than by category (e.g., towards the high end of moderate, low end of high, and so on). For this method, an estimate of metabolic rate is made within the category and a corresponding limit value is obtained by linear interpolation. The third form is where a more accurate measure of metabolic rate is used and provides a continuous form of providing limits (convenient for computer-based systems). The following equation is used for acclimatized people: $WBGT_{ref} = 56.7 - 11.5 \log_{10}(M)$ °C, and for unacclimatized people: $WBGT_{ref} = 59.9 - 14.1 \log_{10}(M)$ °C, where $115 > M < 520$ and is the metabolic rate in Watts. These are also the equations used to specify limits by NISOH (2016).

Annex A of the standard provides the WBGT reference values. It is an informative annex and as such is not compulsory. It therefore represents best practice in the absence of further information. It could be interpreted as, if more experience and information regarding the context of the heat stress exposure were available, then different limits could be integrated into a heat stress management system. It would seem reasonable, however, for the onus to be on the managers to justify any change.

Annex B of the standard is the only normative [compulsory to comply with the standard ("shall" not "should" is used)] annex and presents a specification of instruments for measuring the WBGT. The natural wet-bulb temperature sensor is cylindrical (6 ± 1 mm diameter, 30 ± 5 mm long), with measuring range 5°C–40°C, and accuracy ±0.5°C. The whole of the sensitive part of the sensor shall be covered by a white wick of highly water absorbent material (e.g., cotton) and the support of the sensor shall have the same diameter as that of the sensor and 20 mm of it shall be covered by the wick. The wick shall be kept clean, woven into the shape of a sleeve, and be fitted over the sensor with precision. The lower part of the wick shall be immersed in a reservoir of distilled water, and the free length of the wick in air shall be 20–30 mm. The water in the reservoir shall not heat up by radiation.

The globe temperature sensor is a black sphere (matte black with mean emission coefficient of 0.95) of 150 mm diameter with thickness of material as thin as possible. Copper is often used, but the material type will not affect the reading, only the response time, which presumably is why the material type is not specified. The measuring range is 20°C–120°C, with an accuracy of ±0.5°C up to 50°C and ±1°C from 50°C to 120°C. Natural wet-bulb temperature and globe temperature are termed derived parameters. Air temperature is a basic parameter and can be measured by any suitable method. The sensor shall be protected from radiation while allowing air flow around it, and any shielding device should not reradiate heat to it. The range shall be 10°C–60°C with and accuracy of ±0.5°C.

Annex C provides equations for calculating estimates of globe temperature for the required 150 mm diameter globe, from globe temperatures measured on different diameter globes. Annex D provides equations for calculating an estimate of natural wet-bulb temperature from basic parameters, air temperature, radiant temperature, air velocity, and humidity. Annex E provides a table of metabolic rates by category (see Table 4.1) and Annex F provides CAVs.

GLOBAL WARMING, HEAT STRESS, THE WBGT, AND PRODUCTIVITY

The prediction of global warming naturally leads to suggestions that more people will experience heat stress and to a more extreme level. A reasonable consequence of warming the planet will be more varied and volatile weather, including heat waves. There have been a series of projects that have considered the effects of global warming across the world, and they generally conclude that there will be greater effects on health and productivity, particularly in industrially developing countries and those that have agricultural economies (Levi et al., 2018; Casanueva et al., 2018; Kjellstrom et al., 2018; Asamoah et al., 2018; Gao et al., 2018; Bröde et al., 2018; Lemke et al., 2018; Flouris et al., 2018). The principles of assessing heat stress as described in previous chapters of this book will continue to apply; however, they must be applied to the new challenge.

The WBGT limits in ISO 7243 (2017) can be met if a system of work and rest is used so that the time-weighted average of the WBGT exposure is within the limits. If we consider agricultural work and consider WBGT limits, we find that for acclimatized workers and a metabolic rate of 300 W, the limits are 28°C WBGT. So, if the WBGT in the field was 30°C WBGT, and in a rest area was 25°C WBGT, then a person would have to rest for 24 min every hour. If the temperature increases by 1°C WBGT, then 31°C WBGT in the field, with 25°C WBGT at rest, would mean a rest of 30 min. That is an extra 48 min of rest in an 8-h day with significant drop in productivity (see also Kjellstrom et al., 2009).

Garzón-Villalba et al. (2017, 2018) considered WBGT exposure limits (threshold limit values; ACGIH, 2017) and found that despite limitations of the WBGT index (Budd, 2008; d'Ambrosio Alfano et al., 2014), the limits (similar to those in ISO 7243, 2017) were protective. That is, when considering a large group of subjects exposed to heat stress, for a range of clothing, for exposures below the limits, there were no cases where internal body temperature did not reach an upper stabilized value (sustainable), whereas for heat stress above the WBGT limits, some cases were found to be sustainable and some unsustainable. The limits are those accepted nationally by many countries (e.g., NIOSH, 2016; ACGIH, 2017; BS EN ISO 7243, 2017) by the European Union (EN ISO 7243, 2017) and internationally (ISO 7243, 2017) as those for the avoidance of heat casualties during exposure to heat stress.

That the limits are protective (often referred to as conservative) is a general finding of studies. They are often criticized for not allowing workers to be exposed to heat when they are likely not to sustain heat injury. However, after the subjects have been screened for health (ISO 12894, 2001), limits in a standard should be protective of all people who are exposed. Risk assessment is a separate issue for managers of heat stress where consequences of preventing people from being exposed to heat, and hence from carrying out a task, work shift or mission, for example, may have to be considered.

NATIONAL INSTITUTE FOR OCCUPATIONAL SAFETY AND HEALTH (NIOSH) 2016 REVISED CRITERIA FOR OCCUPATIONAL EXPOSURE TO HEAT AND HOT ENVIRONMENTS

Jacklitsch et al. (2016) in consultation with a team of international scientists updated a previous publication for criteria for a recommended standard on occupational heat stress (NIOSH, 2016, 1986). References are made to an information sheet on heat illness (OSHA, 2011) as well as a publication on the prevention of heat casualties in outdoor workers (NIOSH, 2013) and research into the additional risk of people to heat caused by age, obesity, and pregnancy (Kenny et al., 2010; Gagnon and Kenny, 2011).

The definition of heat stroke has been updated to include two types. These are classical heat stroke and exertional heat stroke. Classical heat stroke is where there is disruption of the central nervous system (collapse or convulsions), a reduction or cessation of sweating, and a rectal temperature of over 41°C. Exertional heat stroke occurs in a physically active person with often continued sweating but failure of skeletal muscle (rhabdomyolysis).

NIOSH (2016) proposes Recommended Alert Limits and Recommended Exposure Limits in terms of WBGT values. A comparison of limits in the United States showed similarities between WBGT limits used by NIOSH, Occupational Safety and Health Administration (OSHA), American Conference of Governmental Industrial Hygienists (ACGIH), and American Industrial Hygiene Association (AIHA). They are also similar to WBGT limits used internationally by the ISO, United Kingdom, Japan, Canada, Sweden, and more. ISO 7243 (2017) limits are identical to those proposed by NIOSH. When the new ISO standard was first proposed, revised limits were considered, but objections from member countries noted that there was no scientific evidence for change and that limits were embedded in the law of some countries. A final proposal accommodated this uncertainty and the limits proposed are similar to those used worldwide. The limits are considered protective of those exposed to heat stress, and there is no scientific reason why standards cannot be harmonized and be identical to the ISO/NIOSH WBGT limits (see ISO 7243, 2017).

WBGT LIMITS IN SPORT

Sport involves people at high levels of activity who are highly motivated to achieve a goal. It comes in many varieties and often takes place in hot environments. Heat illness and death in sport is a serious worldwide problem. Heat is the main cause of illness and death among U.S. high school athletes and is particularly prevalent among non-elite athletes in fun runs and marathons who are fit but not experienced and take longer to complete a race so have longer exposure times. There are many more examples.

The American College of Sports Medicine (ACSM, 1984) provides practical guidance. For people wearing normal running gear (do not run in the heat in a furry animal outfit), risk of heat casualties is very high for WBGT values at the race site of 28°C; high between 23°C and 28°C WBGT; moderate for 18°C–23°C WBGT with less than 18°C WBGT as low. Sports Medicine Australia proposes limits from 20°C WBGT for low risk of heat casualties, to high to very high risk, above 26°C WBGT. If WBGT is 30°C or above, the event should not go ahead. It is important that the WBGT measurements are taken according to the correct specification (ISO 7243, 2017) and that the financial and social consequences of cancelation or postponement do not outweigh considerations of the human cost.

Racinais et al. (2015) provide "Consensus Recommendations on Training and Competing in the Heat." An international panel of experts agreed on advice based upon a review of current proposed heat stress management systems and WBGT limits for a range of sports (see Table 7.1). Consideration is given to a range of preparation and cooling methods, and emphasis is given to acclimatization and hydration as part of preparation for a sporting event in hot conditions. Although considered, WBGT limit values are provided for a range of sports, the evidence for the limits seems sparse. Some of the limits seem at a high level for the very high metabolic rates found in sporting activities. Unless valid data and harmonization of standards and limits takes place across sporting activities (and some military exercises), it is reasonable to expect heat casualties, and deaths due to heat, to continue to occur.

TABLE 7.1 A sample of recommended actions and WBGT limits for a range of sports

WBGT (°C)	*ORGANIZATION*	*ATHLETES*	*RECOMMENDATION*
32.3	ACSM	Acclimatized, fit, low risk	Participation cutoff
32.2	ITF	Junior and wheelchair tennis players	Immediate suspension of play
32.2	WTA	Female tennis players	Immediate suspension of play
32.0	FIFA	Soccer players	Additional cooling break at 30 and 75 min
30.1	ACSM	Nonacclimatized unfit, high risk	Participation cutoff
30.1	ITF-WTA	Junior and female tennis players	10 min break between second and third set
30.1	ITF	Wheelchair tennis players	Suspension of play at the end of the set in progress
28.0	ITF	Wheelchair tennis players	15 min break between second and third set
28.0	Australian Open	Tennis players	10 min break between second and third set
21.0	Marathon in northern latitudes	Runners in mass participation event	Cancel marathon

Source: From Racinais et al. (2015).
ACSM, American College of Sports Medicine; FIFA, Federation Internationale de Football Association; ITF, International Tennis Federation; WTA, Women's Tennis Association.

Assessment of Heat Stress Using the Body Heat Equation

8

EVAPORATION REQUIRED FOR HEAT BALANCE

Heat stress on a person raises the skin temperature through vasodilation of peripheral blood vessels and causes sweating at a level required to lose sufficient heat by evaporation for homeostasis and hence heat balance. That is the required evaporation (E_{req}). The amount of sweat required to achieve this is the required sweat rate (Sw_{req}).

Required sweat rate is the sweat rate required to provide the evaporative cooling sufficient to balance any heat gains by metabolic rate and by convection and radiation from the skin and through breathing. Evaporation required is equal to the metabolic rate (M) minus energy used for mechanical work (W) minus any heat loss by convection and radiation from the skin ($C + R$) minus any heat loss by convection and evaporation by breathing ($C_{res} + E_{res}$). By convention, metabolic energy is positive, and heat transfer symbols are heat losses so a negative heat loss is a heat gain ($E_{req} = M - W - C - R - C_{res} - E_{res}$). If, for example, a person is walking in the sun with a light breeze, then possible values are 180 W metabolic heat production ($M - W$); $C = 10$ W; $R = -40$ W; and $C_{res} + E_{res} = 15$ W, then $E_{req} = 180 - 10 + 40 - 15 = 195$ W. Assuming efficient evaporation from the skin, evaporation of 1 L of sweat in 1 h will transfer 675 W to the environment, so 195/675 = 0.29 L of sweat are required to be evaporated per hour, which means at least that amount of sweat produced, depending upon sweating to evaporation efficiency (influenced by clothing, dripping, humidity, and so on).

In any heat stress environment, heat transfer by all of the terms in the above heat balance equation can be calculated from air temperature, radiant temperature, air velocity, humidity, an estimate of metabolic rate, and by taking account of the heat transfer properties of clothing. This provides us with a rational (analytical) approach to the assessment of heat stress.

The required sweat rate alone will provide us with an indication of thermal strain and is an effective heat stress index (HSI). In all people, there will be a maximum amount of sweat that a person can produce (SW_{max}). If $SW_{req} < SW_{max}$, then SW_{req} is an effective HSI. If $SW_{req} > SW_{max}$, then the person will sweat at a maximum level, but there will be insufficient heat loss by evaporation and there will be heat storage (S) and a consequent rise in body temperature. The rate of heat storage (increased body temperature) will depend upon the difference between SW_{max} and SW_{req}. In fact, heat storage (body temperature rise) will depend upon the difference between E_{max} and E_{req} as E_{max} will also depend upon clothing and the environmental conditions. Exposure limits can therefore be calculated based upon predicted body core temperature (e.g., how long it will take to reach a core temperature of 38°C). They can also be based upon the level of dehydration as sweating will cause water loss.

There have been various forms of rational HSI proposed. All are based upon the principles described previously. An early version was by Belding and Hatch (1955). They proposed the HSI as required evaporation divided by the maximum evaporation that can occur in the environment (HSI = E_{req}/E_{max}). Gagge (1937) had proposed a similar index called skin wettedness (w), which is the amount of heat loss by evaporation divided by the maximum amount that could be lost if the whole of the skin were wet with sweat. Values ranged from $w = 0.06$ for no sweating to $w = 1$ for the skin completely wet. This raised the concept of required skin wettedness (w_{req}).

Givoni (1963, 1976) proposed the Index of Thermal Strain as an extension of the HSI to include an explicit solar radiation constant and to recognize that mechanical work as well as metabolic heat production make up metabolic rate. Efficiency of sweating is also included, recognizing that not all sweat evaporates.

Applied research for the iron and steel community of the European Union (Metz, 1988) developed a HSI based upon the required sweat rate (Sw_{req}). Vogt et al. (1981) proposed the index which calculated the required sweat rate based upon an improved heat balance equation and also a method of interpretation by comparing what is required with what is physiologically possible and acceptable in workers. After much discussion and evaluation, the Sw_{req} index was adopted internationally by ISO 7933 (1989). The Sw_{req} value is calculated from a heat balance equation using the six basic parameters and posture [effective radiation surface area (0.72, sitting; 0.77, standing)]. Required sweat rate is given as $Sw_{req} = E_{req}/r$ where r is sweating efficiency. Required skin wettedness $w_{req} = E_{req}/E_{max}$.

Based upon reference values of what is acceptable and what people can achieve, predictions of skin wettedness (w_p), evaporation rate (E_p), and sweat rate (Sw_p) are derived. If what is required can be achieved, these are the predicted values and it is concluded that the work can be continued for an 8-h shift as long as workers do not lose more than 4% (Warning) and 6% (Danger) of body weight and hence dehydration. If the predicted values cannot be achieved, maximum values become the predicted values and duration limited exposure (DLE) times are calculated based upon predicted sweat rate and maximum allowable dehydration.

ISO 7933 (1989) provided Sw_{req} as a HSI but was used mainly for its outcome of allowable exposure times (DLEs) for warning and danger levels for acclimatized and unacclimatized workers. As the criteria used were single figure limits (If value is >limit then response changes) discontinuities occurred between environmental conditions and outcomes (Kampmann and Piekarski, 2000). To improve this standard, therefore, a European research program (BIOMED), involving researchers from Belgium, Italy, Germany, the Netherlands, Sweden, and the United Kingdom, developed improvements leading to a proposed, and an extensive evaluation of, a new Sw_{req} method called predicted heat strain (PHS) (Malchaire et al., 2000, 2001). Improvements included consideration of respiratory heat loss, mean body temperature, distribution of heat storage in the body, prediction of rectal temperature, exponential averaging for mean skin temperature and sweat rate; evaporative efficiency of sweating; w_{max} limits for nonacclimatized subjects; maximum sweat rate; increased core temperature with activity; limits of internal temperature; maximum dehydration and water loss; influence of radiation on clothing; and the inclusion of the effects of ventilation on clothing insulation. The new method was published as ISO 7933 (2004).

ISO 7933 (2004): ERGONOMICS OF THE THERMAL ENVIRONMENT—ANALYTICAL DETERMINATION AND INTERPRETATION OF HEAT STRESS USING CALCULATION OF THE PHS

This international standard uses the heat balance equation to predict the sweat rate and internal body temperature of people in response to heat stress. It can also be used to identify which parameters or group of parameters could be modified to reduce heat strain. It provides the evaluation of heat stress likely to lead to excessive core temperature and water loss and a method for determining maximum allowable exposure times within which no physical damage is to be expected. It considers standard subjects in good health and fit for the work they perform.

A slightly modified heat balance equation is used to calculate required evaporation: $E_{req} = M - W - C - R - C_{res} - E_{res} - dS_{eq}$, where dS_{eq} is the body heat storage rate for the increase of core temperature associated with the metabolic rate (see Chapter 2).

Two criteria of thermal stress (w_{max}, Sw_{max}) and two for thermal strain (rectal temperature, dehydration) are used to determine exposure limits. For unacclimatized/acclimatized people, values are for maximum wettedness: w_{max}, 0.85/1.0 (ND) and for maximum sweat rate: Sw_{max}, $[(M - 32) \times A_D]/[1.25 \times (M - 32) \times A_D]$ in $W\,m^{-2}$. For dehydration and rectal temperature, limits are identical for unacclimatized and acclimatized people. Maximum dehydration and water loss when people can drink freely is for 50% of the population protected: D_{max50} = 7.5% of body mass and for 95% of the population protected: D_{max95} = 5% of body mass (alarm level). For people who cannot drink freely, the limit is 3% of body mass loss. For rectal temperature, a limit of 38°C is used.

The sweat rate in watts per square meter represents the equivalent in heat of the sweat rate expressed in grams of sweat per square meter of skin surface per hour where $1\,W\,m^{-2}$ corresponds to a flow of $1.47\,g\,m^{-2}h^{-1}$ or $2.67\,g\,h^{-1}$ for a standard subject ($1.8\,m^2$ body surface area) (ISO 7933, 2004). Heat storage and rise in rectal temperature assumes a specific heat of the body of $3.49\,kJ\,kg\,K^{-1}$ and takes account of the shell/core distribution. As the body heats up, there is a greater proportion of core.

The hot environment is assessed by measuring the environmental parameters (T_A, T_R, RH, v) and estimating values of metabolic rate and of the thermal properties of clothing. The evaporation required (E_{req}), the skin wettedness required (w_{req}), and the sweat rate required (Sw_{req}) are calculated from the heat balance equation and predicted evaporative heat flow (E_p), skin wettedness (w_p), and sweat rate (Sw_p) are derived, taking account of the physiological limitations of the body (Sw_{max}, D_{max}) and the exponential response of the sweating system. The difference between E_{req} and E_p provides heat storage (S) which leads to predictions of mean skin temperature (T_{SK}) and rectal temperature (T_{RE}). An incremental time system is used such that, as heat stress parameters vary, the Sw_p and T_{RE} variation with time is computed.

Maximum allowable exposure time (D_{lim}) is reached when rectal temperature or cumulative water loss reaches limit values. If condensation occurs or exposure time is less than 30 min, then the standard is not valid, and physiological monitoring of people exposed to the heat is recommended (ISO 9886, 2004). A computer program and example input and results data are provided. The method was validated using a database of human responses to heat, including 347 laboratory experiments and 366 "field" experiments from eight research institutions. A valid range of parameter values is provided. Of particular note is the valid range for clothing (0.1–1.0 Clo). Methods for estimating metabolic rate and clothing properties are provided in annexes to the standard.

Suppose an acclimatized person works in a standing position conducting agricultural work of moderate intensity in the sun, wearing a T-shirt, and light jeans. Air temperature is 30°C; mean radiant temperature 50°C; air velocity 0.3 ms^{-1}; humidity 3.0 kPa (70% relative humidity); metabolic rate is estimated to be 150 $W m^{-2}$ (270 W); and clothing insulation 0.5 Clo.

The predicted final rectal temperature over an 8-h working day is 37.7°C, and 7.166 L is lost as sweat. The interpretation according to ISO 7933 (2004) is therefore that, although sufficient heat can be lost by evaporation to maintain rectal temperature below the limit value of 38°C, assuming that water is freely available to drink, only 50% of the workers would be protected from dehydration after 380 min of work and 95% of the workers would be protected after 258 min which is the "alarm" level. Action is required to reduce the heat stress (e.g., provide shade), restrict exposure times (work/rest cycles and so on), or implement other practices according to context. The new conditions and arrangements would be advised and re-evaluated using the standard.

ISO 7933 (2004) COMPUTER PROGRAM

Annex E to ISO 7933 (2004) provides a full program written in the computer language Quick Basic. This is described with complete information such that the reader can construct their own computer program. The reader is left to use whichever software platform and interface they wish to use. For versions freely available on the Internet, the standard refers to a useful website. The reader should confirm their accuracy with the test data that is supplied in ISO 7933 (2004). A fuller description is provided in Parsons (2014), and for practical application, the reader is referred to the standard which also includes a flow diagram of the PHS method.

ISO DIS 7933 (2018)

The main objective of this proposed new international standard will be to describe a mathematical model (the PHS model) for the analytical determination and interpretation of the thermal stress (in terms of water loss and core temperature) experienced by a subject in a hot environment and to determine the "maximum allowable exposure times," with which the physiological strain is acceptable for 95% of the exposed population (i.e., the maximum tolerable core temperature and the maximum tolerable water loss are not exceeded by 95% of the exposed people).

ISO 7933 (2004) is not applicable to cases where special protective clothing (reflective clothing, clothing with active cooling and ventilation, impermeable clothing, and clothing when wearing personal protective equipment) is worn. Also, it does not account for transients in environmental conditions, metabolic rate, or clothing. The proposed new version of the standard will also be limited in these areas.

One of the main proposed changes to ISO 7933 (2004) is in the method of estimating metabolic rate. It is proposed that while a simple estimation from a screening method is acceptable for ISO 7243 (2017), higher accuracy is required for this analytical method. In particular, the use of heart rate as a method for estimating metabolic rate is recommended and guidance is provided on how this can be achieved.

The human heat balance equation provides a pragmatic analytical method for assessing human heat stress. Its formulation so far has been to concentrate on heat transfer mechanisms and pay little attention to the time varying nature of heat stress and the heat strain it elicits and on the thermoregulatory mechanisms of the body (Chapter 2). Simple, often discontinuous, and empirical methods are used (e.g., regression equations to provide best fit). Elaboration of rational methods based upon a simple heat balance equation have probably gone further in complexity than the application requires and has probably detracted from usability and hence use. This accounts for the success of the WBGT index. Paradoxically, a more comprehensive approach involving a simulation of the person in hot conditions, including heat transfer, a dynamic system of thermoregulation and a passive system that includes the thermal properties of tissues and more, can be more usable and at least as valid. These are called thermal models and are described in Chapter 9.

Computer Models for Assessing Human Heat Stress

9

HUMAN MODELS AND HEAT STRESS

A 38-year-old Texas murderer, a prisoner on death row, was executed by lethal injection on August 5th, 1993. He had donated his body to science, and after his death, it was sliced very thinly so that a cross section of all the tissues and their distribution and dimension was obtained as well as a wealth of other information albeit from one dead individual. The data, followed by others, including the body of a female housewife, were placed on the Internet for all to have access (Waldby, 2000).

Studies in human anatomy, anthropometry, and experimental physiology allow us to build a model of the human body in both dimension and construction. We also have an idea of the proportion, distribution, and thermal properties of its tissues. Unlike Frankenstein's monster, we do not have to build this model, or dead body (although human manikins and robots are used in clothing assessment and other applications), but we can describe it mathematically. As soon as the computer became viable, it became possible to define a passive human body. Studies of the metabolic rate provided information about heat generation within the body, and control theory also lent itself to computer representation. So not only could we build a passive body but we could install a control system of thermoregulation to bring it to life. It was then possible to study how the model, and hence maybe a human, would respond to any thermal environment including heat stress.

COMPUTER MODELS OF HUMAN THERMOREGULATION

The first models of human thermoregulation, as they were called, emerged in the 1950s, mainly with work in the U.S. military. Building on the work of Machle and Hatch (1947) who introduced the terms core and shell temperature, and the heat perfusion work of Pennes (1948), Wissler (1961) produced a model of the human body, based upon a single cylinder, for the U.S. Air force. Stolwijk and Hardy (1966) based at the J. B. Pierce laboratories in Yale, USA, produced a model of human thermoregulation for NASA (National Aeronautics and Space Administration) and along with colleague Gagge as well as others, went on to develop the Stolwijk and Hardy model which has formed the basis of most models of human thermoregulation (see Stowijk and Hardy, 1977).

Until the late 1970s to early 1980s, computers had been the domain for the specialist and the scientist and engineer. Computer programs were mainly procedural, used computer languages such as FORTRAN, and often each instruction was printed on a computer card. Computer models of human thermoregulation were therefore often manifest as a large box of computer cards. Computers were relatively slow, thus, to conduct a simulation involving numerous calculations often took 24 h or more. As the computer developed, "finally" onto the Internet, this ceased to be a limitation.

THE STOLWIJK AND HARDY MODEL

The Stolwijk and Hardy (1977) computer model of human thermoregulation was called the 25-node model. It is termed a lumped parameter model as it is considered as 25 entities (lumps!). The passive system of the model represents the body as six body areas (segments). These are the head, which is represented as a sphere and the trunk, arms, hands, legs, and feet, which are represented as appropriately sized cylinders. Each segment has four layers, core, fat, muscle, and skin. All of the segments are perfused with blood which is the 25th node (6 segments × 4 layers + 1 blood = 25 nodes). All of the nodes are defined completely in terms of their dimensions, mass, thermal properties (specific heat capacity and so on), basal metabolic heat, and so on (see Stolwijk and Hardy, 1977; Haslam and Parsons, 1988; Parsons, 2014).

The model is driven by a controlling system. This is a thermoregulatory system that is based upon set points which are provided for each layer and thermoregulatory activity. If the set point for sweating is exceeded, for example, then the skin evaporation increases (distributed across the skin of segments in a way similar to that of the human body). Vasodilation increases the skin temperature by a similar mechanism, and activity level producing metabolic heat is distributed across mainly muscle layers. Heat transfer inside and between segments (via blood flow) and through clothing to the outside environment (defined by air temperature, radiant temperature, air velocity, and humidity) or through clothing from the outside environment, provide a dynamic interaction. The computer program begins at a starting condition for all nodes, environment, and thermoregulatory position (e.g., thermally neutral and comfortable), and in increments (e.g., of 1 min or less) progressive states of the body are computed. This gives a prediction of the physiological state of the human body with time, for all nodes (segments and layers) as well as heat transfer information and more.

THE 2-NODE MODEL OF HUMAN THERMOREGULATION

Gagge et al. (1971), proposed a 2-node model of human thermoregulation, initially concerned with developing a thermal comfort index, Nishi and Gagge (1977) but with potential for use in the assessment of heat stress. The two nodes are body core and shell in concentric cylinders. The model is a simple version of more complex models and is presented below as a demonstration of how a computer can provide a simulation of human response to the thermal environment.

The following description of the 2-node model of human thermoregulation, inspired by Pharo Gagge and others at the J B Pierce Foundation Laboratory, Yale University, USA, is a modified version of the model, written in FORTRAN, presented by Haslam and Parsons (1989a). The intention is to provide sufficient information for the reader to construct their own model and also to demonstrate the construction, nature, and viability of human thermal computer models in general, using this simple example. To link with, and get a feel for, the computer program, some of the variables and code used are provided as they appear in the computer program. The variables are labeled in a way that indicates what they represent such that the reader will be able to follow the code (e.g., TCR is core temperature, TSK is skin temperature, and so on).

THE PASSIVE SYSTEM, INPUTS, AND A THERMONEUTRAL START

A simulation of a person in a thermal environment is continuous and varies in time. A model therefore requires starting conditions. These can be defined by the model. In fact, it is the utility of a model that we can start a simulation with the person hot, cold, neutral, or in between, they could be fat, thin, muscular, tall, short, male, female, child, baby, have an illness, or disability and wear a swimsuit, sports, or protective clothing and more. We could also define the environment as hot, cold, comfortable, at sea level, up a mountain, in water, underwater, in pressurized air, and in space. All will depend upon the parameters available in the model. A continued simulation can start with the person's condition as at the end of a previous simulation, so providing responses to changing environments. For the 2-node model, it is usual to start (at time = 0) with a standard healthy male, exposed to conditions of interest (e.g., heat stress) in the starting physiological state of thermal neutrality. The "passive" part of the body is represented by two concentric cylinders (core inside a shell), and clothing is represented as thermal resistance to heat transfer but without mass (inertia). The point is that all parameters in the model can be predefined and changed as required in the simulation. For the purposes of the following example, a steady-state model with minimum inertia is assumed. Inertia (e.g., mass of clothing) will cause lag when conditions change and can be added along with other properties if they become relevant to the simulation.

The computer program which implements the model requires inputs and results, supplied through its user interface. In the current example, these are air temperature (TA, °C), mean radiant temperature (TR, °C), air velocity (V, $\mathrm{ms^{-1}}$), relative humidity (RH, fraction), the thermal insulation of clothing (CLO, clo), the intrinsic vapor permeation of clothing (IECL, $\mathrm{m^2\,kPa\,W^{-1}}$), total metabolic rate (MR, $\mathrm{W\,m^{-2}}$), and external work (WK, $\mathrm{W\,m^{-2}}$). The starting core temperature (of the inner cylinder) is 36.8°C and skin temperature (of the outer cylinder) is 33.7°C. For the results, although calculations are made for every minute of exposure, the output interval time (ITIME) is provided (e.g., every 15 min) as is the total exposure time (TTIME; e.g., 8 h). So for example, we can predict core temperature and skin temperature and more and display the values every 15 min over an 8-h exposure to any environment as defined by the inputs.

The starting conditions of the model described are mostly from Gagge et al. (1986). These are skin (shell) temperature (TTSK = 33.7°C); core temperature (TTCR = 36.8°C); skin wettedness (PWET = 0.06); metabolic rate [MR = WORK (inputted earlier)]; the thermoregulatory controller for sweating (CSW = 170); the controller for vasodilation (CDIL = 200); and the controller for vasoconstriction (CSTR = 0.1). The controller constants relate to how sensitive the model is in terms of the level of response compared with a level of change in conditions. As the body becomes hot and vasodilation occurs, more of the body can be considered as core and less as shell (less core, more shell, for vasoconstriction and cold). The term ALPHA is used to represent the fraction of the body which is shell. ALPHA = 0.1 is the thermoneutral starting point which gives a starting mean body temperature TTBM = $0.1 \times 33.7 + 0.9 \times 36.8 = 36.49$°C. As the condition of the body varies with time, the mean body temperature (TBM) varies with ALPHA, mean skin temperature (TSK), and core temperature (TCR). Skin blood flow (SKBF) starts at a value of 6.3. These values are constants that make the model simulate a person in thermal neutrality.

THE THERMOREGULATORY CONTROLLER

The thermoregulatory controller is driven by the difference between the thermoneutral (starting) values for skin and core temperatures, and the calculated values after exposure time increments, based upon heat storage. If the temperatures rise, blood flows to the skin and a greater proportion of the body becomes core (1—ALPHA). Sweating is controlled by the difference between calculated mean body temperature and neutral mean body temperature. Vasoconstriction is controlled in a similar way to vasodilation, but the proportion of shell (ALPHA) is increased. To simulate shivering in the cold, metabolic rate is increased as a function of changes in both the skin and the core temperatures.

TIMERS TO PLOT VARIATIONS IN BODY CONDITION WITH TIME

All counting in the computer code is set to minutes. The body condition is set to its thermoneutral starting point, and calculations of its future states are made at intervals (e.g., every minute) up to a point where we wish to see the output (e.g., every minute for 8 h may be too cumbersome so we may view only every 15 min). This is repeated then concluded when the total exposure time is reached.

The following times and timers are therefore used in the computer program. Actual time (MTIME in min, from 0 to total exposure time); increment time (NTIME = 1 min); increment time (DTIME in h); total exposure time (JTIME in min); total exposure time (TTIME in h; JTIME = 60 × TTIME); ITIME (in min); and output interval timer (KTIME in min).

There are two timers (MTIME and KTIME). IF MTIME >= KTIME the results are "printed.". MTIME is incremented by NTIME and KTIME is incremented by ITIME. Starting values are MTIME = O; NTIME = 1; DTIME = 1./60; and JTIME = 0. "Loops" in the computer program and calculations at relevant positions provide the output condition of the body [e.g., indicated by selected parameters such as core and skin (shell) temperatures, heat transfers and so on] at selected intervals, and over total exposure time. For example: If the exposure time has not been reached, go to the start of the thermoregulatory loop, otherwise output the final results. IF (MTIME.LT.JTIME) GO TO 500. Where 500 CONTINUE is the label at the start of the thermoregulatory loop. To aid the reader in interpreting the computer code, we should note that when the symbol "=" is used we should read it as "becomes." * means multiply by; ** means to the power of; >= or GE means greater than or equal to (LT means less than and so on). Real numbers and integers may also have to be identified (e.g., 4. to indicate a real number).

GAGGE'S 2-NODE MODEL—THE COMPUTER PROGRAM

Having set variables, inputs, and outputs and defined the passive system, the computer program starts with a labeled code line (e.g., 60 CONTINUE). This allows us to go back to the beginning in a loop, a mechanism used throughout this program. The timers are set for KTIME (min) and exposure time JTIME (min): KTIME = KTIME + ITIME: JTIME = JTIME + TTIME*60

The temperature regulation loop starts with a calculation of heat transfer and determination of heat storage over the time increment (time = 0, then first minute, and so on) using the body heat equation.

The convective heat exchange coefficient (CHC) is for still air (natural convection) CHC = 3.0 and for forced convection (CHC>3.0), CHCV = 8.6*V**0.53.

The heat transfer radiation coefficient (CHR) and mean surface temperature of clothing (TCL) are determined in an iterative loop starting (below) with TCL = 0 (for later convenience evaporation at the skin, ESK = 0 is also included here).

600 TCLOLD = TCL updates the estimate of clothing surface temperature

CHR = 4.*5.67E-8*((TCL + TR)/2. + 273.2)**3)*0.725 calculates an estimate of the coefficient of the heat transfer by radiation from the estimate of surface temperature of clothing.

TCL = (1./(CLO*0.155)*TSK + FACL*(CHC*TA + CHR*TR))/((1./CLO*0.155) + FACL*(CHC+CHR)) calculates the estimate of TCL using the estimate of CHR.

IF (ABS(TCL – TCOLD).GT.0.01) GOTO 600 sends the calculation back to the beginning as both equations are not satisfied. If the calculated TCL has not changed significantly (change is less than or equal to 0.01), the equations are satisfied, and we have TCL and CHR values.

Respiratory heat losses are:

```
ERES = 0.17251*MR*(5.8662-RH*SVP(TA))
CRES = 0.0014*MR*(34-TA)
```

SVP(TA) is the saturated vapor pressure at air temperature given as a function of any temperature T as SVP(T) = 0.133322*EXP(18.6686 – 4030.183)/(T + 235).

Heat flows are given as

```
DRY = FACL*((CHC*(TCL-TA) + CHR*(TCL-TR)
```

Heat flow at the skin

```
HFSK = (TCR-TSK)*(5.28 + 1.163*SKBF) - DRY - ESK
```

Heat flow from core

```
HFCR = MR - (TCR - TSK)*(5.28 + 1.163*SKBF) - CRES -
ERES - WK
```

For the average man used in the example (70 kg, 1.7 m, 1.8 m^2), TSK and TCR are calculated over the time increment (first minute) as

```
TSK = 0.97*ALPHA*70: TCCR = 0.97*(1 - ALPA)*70
DTSK = (HFSK*1.8)/TCSK: DTCR = (HFCR*1.8)/TCCR
TSK = TSK + DTSK*DTIME: TCR = TCR + DTCR*DTIME
```

Thermoregulatory signals (simulating signals from skin and core to the brain to elicit a response) are driven by the difference between calculated values for TCR and TSK and the starting thermoneutral values.

Control signals for blood flow

```
SKSIG = TSK - TTSK: IF(SKSIG.LT.0): COLDS = -SKSIG,
WARMS = 0: ELSE COLDS = 0, WARMS = SKSIG
CRSIG = TCR - TTCR: IF(CRSIG.LT.0): COLDC = -CRSIG,
WARMC = 0: ELSE COLDC = 0, WARMC = CRSIG
```

Control of skin blood flow (SKBF limited from 0.5 to 90)

```
STRIC = CSTR*COLDS: DILAT = CDIL*WARMC: SKBF = (SKBFN +
DILAT)/(1. + STRIC)
ALPHA = 0.0417737 + 0.7451832/(SKBF + 0.585417)
```

Control signals for sweating

```
TBM = ALPHA*TSK + (1 - ALPHA)*TCR: BYSIG = TBM - TTBM
IF (BYSIG.LE.0.) COLDB = -BYSIG, WARMB= 0.: ELSE WARMB =
BYSIG, COLDB = 0.
```

Control of regulatory sweating

```
REGSW = CWS*WARMB*EXP(WARMS/10.7): IF (REGSW.GT.500)
REGSW = 500.0
ERSW = 0.68*REGSW
```

Heat transfer by evaporation

```
LR = 15.1512*(TCL + 273.15)/273.15: FACL = 1 + 0.31*CLO:
RT = IECL + 1./(FACL*LR*CHC): EMAX = (1./RT)*(SVP(TSK) -
RH*SVP(TA))
PRSW = ERSW/EMAX
PDIF = (1.0-PRSW) * 0.06: EDIF = PDIF*EMAX: ESK = ERSW+
EDIF: PWET = ESK/EMAX
```

Note for shivering: MR = WORK + 19.4*COLDS*COLDC

Total evaporative heat loss: EV = ESK + ERES and total dry heat loss: DRYT = DRY + CRES. MTIME is incremented by 1 minute (NTIME).

IF (MTIME.GE.KTIME) output results and set KTIME=KTIME+ITIME

IF (MTIME.LT.JTIME) GOTO 500 so the regulatory loop continues until total exposure time is reached. Heat storage and skin heat loss are calculated as STORE = MR – WK – CRES – EV – DRY: HSK = MR – ERES – CRES – WK – STORE.

Outputs

The simulation can produce at least as many outputs as there are parameters in the model and calculations determined from them. Useful outputs can be selected in the context of the application. If we are conducting research into human thermoregulation then controller outputs may be of interest. In the context of heat stress, core temperature and water loss will be of interest. The time progression ($t=0$ to exposure time) of core temperature, skin temperature, and skin wettedness will be primary indicators of thermal strain and metabolic rate, dry heat transfer ($C+R$), total evaporative heat loss, and evaporative loss at the skin as well as total water loss will be of interest. Thermal models can therefore be used to assess heat stress and provide a direct determination of thermal strain. Interpretation may include exposure limits based upon physiological condition (core temperature or water loss) and make the use of a heat stress index for that purpose, redundant.

Using information provided in this chapter, a computer program was written in the programming language Visual BASIC with a spreadsheet interface. The inputs and outputs are shown in Figure 9.1 for the model simulation of a person under heat stress.

Figure 9.1 provides an example of a simulation of the responses of a person, slowly walking for 1 h, in an environment of 40°C, air and mean radiant temperature, 60% rh, with still air and light cotton clothing. The person starts in a comfortable condition, and core temperature steadily rises to 38.26°C (models provide as many digits after the decimal point as required but that should not be confused with accuracy). The skin temperature converges with the core temperature from 33.7°C to 37.8°C due to vasodilation. Dry heat loss by convection and radiation ($C+R$) is negative so indicates a gain (which makes sense as the air and mean radiant temperatures are greater than body temperatures), which reduces in time as skin temperature rises. Evaporative heat loss increases with sweating up to a maximum. Maybe more quickly at first than in an actual person, as in practice, there will be a lag. As sweating increases, it provides more heat loss by evaporation to counteract the heat

				INPUTS				
Air Temperature	Mean Radiant Temperature	Air Speed	Relative humidity	Intrinsic clothing insulation	Clothing area factor	Intrinsic Clothing vapour permeability	Initial metabolic rate	External work
TA°C	TR°C	V ms-1	Fraction (ND)	ICL (Clo)	FACL (ND)	m2 kPa W-1	MR (Wm-2)	Wm-2
40	40	0.1	0.6	1	1.31	0.024	100	0

				OUTPUTS					
TIME	CORE TEMPERATURE	SKIN TEMPERATURE	SKIN WETTEDNESS	METABOLIC RATE	DRY HEAT LOSS (C+R)	TOTAL EVAPORATIVE HEAT LOSS(E)	EVAPORATIVE HEAT LOSS AT THE SKIN	DRIPPING	SKIN TO CORE RATIO
min	°C	°C	(ND)	Wm-2	Wm-2	Wm-2	Wm-2	g/h.m2	(ND)
0	36.8	33.7	0.06	100	0	0	0.00	0	0.07
5	36.9	35.31	1	100	-22.68	38.29	35.81	7.26	0.07
10	37.01	36.33	1	100	-17.48	47.46	44.97	39.32	0.06
15	37.15	36.72	1	100	-15.45	51.05	48.56	71.9	0.05
20	37.28	36.92	1	100	-14.41	52.96	50.48	103.86	0.05
25	37.42	37.04	1	100	-13.89	54.1	51.62	135.34	0.05
30	37.55	37.16	1	100	-13.37	55.22	52.73	166.66	0.05
35	37.68	37.27	1	100	-12.87	56.31	53.82	197.77	0.05
40	37.80	37.38	1	100	-12.39	57.37	54.89	228.68	0.05
45	37.92	37.49	1	100	-11.91	58.41	55.93	259.34	0.05
50	38.04	37.6	1	100	-11.45	59.43	56.95	289.74	0.05
55	38.15	37.7	1	100	-11.01	60.42	57.94	319.87	0.05
60	38.26	37.8	1	100	-10.57	61.39	58.91	349.7	0.05

FIGURE 9.1 Example inputs and results of a simulation of a person under heat stress using the 2-Node model of human thermo-regulation (Adapted from Haslam and Parsons 1989a).

gains by metabolic rate (100 W) and dry heat gain (environmental temperature greater than body temperature). So rate of heat storage slows down until maximum evaporation is reached. The person is soaked with sweat ($w = 1$), and it can be concluded that there is little danger of heat illness for a 1-h exposure. An extended exposure may start to cause problems due to dehydration and maybe increased core temperature. A simulation with greater exposure time would verify when problems may occur. It would also demonstrate the effects of reducing the heat stress, by reducing clothing levels, for example.

OTHER THERMAL MODELS

Wissler (1961) described a six-cylinder model that evolved over a number of years into a model that computed 225 temperatures in 15 elements and included O_2, CO_2, and lactate concentrations (Wissler, 1985). The model of Stolwijk and Hardy (1977) is the model most used as a basis for the development of other models (Parsons, 2014). A series of models was developed for practical use in a study reported by Haslam (1987) and Haslam and Parsons (1988, 1989a,b). Four main models were investigated to see if they were of practical use to predict the thermal responses of soldiers in the British army. These were models that used a heat balance equation to calculate the required sweat rate (Sw_{req}—ISO DIS 7933, 1987); modified (labeled LUT for Loughborough University of Technology) versions of the J B Pierce 2-node model and the Stolwijk and Hardy 25-node model; and the Givoni and Goldman's heat stress prediction models used by the U.S. Army (Givoni and Goldman, 1972). A comparison of predictions of the models with actual responses of people provided an evaluation (Haslam and Parsons, 1989b). Fiala (1998) describes a model that has 19 compartments and 325 nodes and takes account of asymmetry and heat transfer for each compartment. The passive system of the models has head, face, neck, shoulders, arms, hands, thorax, abdomen, legs, and feet. It also includes brain, lung, bone, muscle, viscera, fat, and two layers of skin. This model was modified to provide the basis of the Universal Thermal Climate Index (UTCI—Fiala et al., 2012). Increased computer power and software have led to numerous developments across the world including the use of finite element and finite difference techniques (e.g., Neal, 1998; Zhu, 2001; Tanabe et al., 2002). Parsons (2014) provides a review as well as a description of the evaluation of models in terms of their usefulness and accuracy for practical use.

THERMAL MODELS AND HEAT STRESS INDICES

Two models in particular have been used to support the development of thermal indices with universal application including their use for the assessment of heat stress. The 2-node model of thermoregulation (Gagge et al., 1971) provides a rational basis for the New Effective Temperature thermal comfort index (ET*), later developed into a universal index called the standard effective temperature (SET) index. The UTCI used the modified Fiala model in an attempt to provide a weather and climate reporting index that could be used worldwide (Jendritzky et al., 2012).

The ET* is defined as the temperature of a standard environment (with 50% relative humidity, still air, and mean radiant temperature equal to air temperature) that would give equivalent effect (e.g., comfort) as the actual environment of interest (which may vary in humidity, air temperature, radiant temperature, and air velocity). The equivalent effect is defined as the same skin wettedness (*w*), mean skin temperature (TSK), and heat loss at the skin (HSK). To achieve this, the *w*, TSK, and HSK values are calculated for the actual environment using the 2-node model, and the 2-node model is then used to find the air temperature of the standard environment that would give the same values. For the ET*, light clothing and office work are assumed. For the SET, the definition is extended to add in the same level of clothing (with a slight adjustment for activity level) and the same activity as in the actual environment. The value is then found using the 2-node model as described for ET*. An SET scale is provided that indicates 26.5°C–30.0°C SET as slight sweating, vasodilation, slightly warm to 37.5°C SET as profuse sweating above which would be "very very hot" and lead to failure of thermoregulation.

Over 45 scientists from 23 countries contributed to the development of the UTCI (Jendritzky and de Dear, 2012). For the index, the Fiala model was reduced to 12 compartments and 187 tissue nodes and improved in the area of clothing, heat exchange, and physiological response. The UTCI was defined as the temperature of a standard environment that gave equivalent physiological response. The standard environment was defined as air temperature equal to radiant temperature, meteorological air velocity (anemometer on a 10 m pole) as 0.5 ms^{-1}, 50% relative humidity or 20 hPa if air temperature was above 29°C, clothing by an equation, and activity at 4 km h^{-1} or 135 W m^{-2}. The model was used to provide a polynomial regression equation that provided a prediction of equivalence taking account of the rectal temperature, mean skin temperature, facial temperature, skin blood flow, sweat rate, skin wettedness, and shivering. This can be compared with the simple definition of equivalence used by the 2-node model for the calculation of the SET (Gagge et al., 1973). As the models become more complex, it is inevitable that defining equivalence will become more difficult as equivalence is a concept and for people there is no unique solution. The UTCI gives an indicator of the effect from extreme cold stress to extreme heat stress (Jendritzky and de Dear, 2012; Parsons, 2014).

Human Heat Illness and Prevention

10

HEAT ILLNESS MOSTLY AFFECTS THE VULNERABLE AND THE SPORTY

Heat stress causes thermoregulatory responses and places an additional strain on physiological systems and organs, particularly the heart, which is why it is mainly elderly and vulnerable people who die or are admitted to hospital during heat waves. Physiological systems among vulnerable people may not have the capacity to withstand or sustain the requirements of thermoregulation, but their bodies will attempt to do so. Even before recognized, heat illness appears, such as heat cramps, heat exhaustion, and heatstroke, a person may be in danger. The reason for placing emphasis on vulnerable people and thermoregulatory strain is that most studies and classification systems of heat illness and death have been on fit, young, otherwise healthy people conducting sports or military activity. Most illnesses and deaths, however, occur among vulnerable people in heat waves.

Heat illnesses (sometimes called heat disorders) can be caused by a strain on the heart and circulatory system, sweating, skin disorders, and disturbance of water and electrolyte balance. They can also be caused by an increase in body temperatures that affect cellular and hence bodily function including that of the vital organs. For a detailed description of heat illnesses see Leithead and Lind (1964), WHO (1969), McIntyre (1980), Goldman (1988), Hubbard and Armstrong (1988), Parsons (2014), and NIOSH (2016).

Heat illness ranges from mild cardiovascular and central nervous system disturbance (hypotension and fainting) to evidence of profound cellular damage of the brain, kidneys, liver, and blood clotting mechanisms (heatstroke). The three most prominent illnesses are heat cramps, heat exhaustion, and heatstroke.

Hubbard and Armstrong (1988) note that electrolyte imbalance, hypohydration, exhaustive exercise, acidosis, lack of acclimatization, and drugs are all host factors for heat illness. They propose that more focus should be placed on the effects of heat on the cell and the consequent "leakiness" of the cell membrane and its resultant effects on energy consumption, heat production, and exhaustion. They point out that during hyperthermia, body core temperature often continues to rise, even after collapse, hence, quick recognition and speedy treatment are required. If hyperthermia has occurred, then Shibolet et al. (1962) have noted two courses leading to death. The first is a rapid death caused by an acute and general de-arrangement of biochemical reactions. The second is collapse followed by an apparent recovery including a lucid period followed by degeneration and death with multiple hemorrhage to vital organs.

Goldman (1988) and NIOSH (2016) present comprehensive tables of heat illnesses. The information and some additions and comments are provided in the following section.

TEMPERATURE REGULATION HEATSTROKE

Clinical features of heatstroke include hot dry skin usually red, mottled, or cyanotic with a rectal temperature of over 40.5°C (104°F) leading to confusion, loss of consciousness, and convulsions. With continued exposure, rectal temperature continues to rise and is fatal if treatment is delayed.

Predisposing factors are sustained exertion in heat by unacclimatized workers, lack of physical fitness and obesity, recent alcohol or drug intake, dehydration, individual susceptibility, and chronic cardiovascular disease. The underlying physiological disturbances are failure of the central drive for sweating (cause unknown) leading to loss of evaporative cooling and an uncontrolled accelerating rise in rectal temperature. There may be partial rather than complete failure of sweating. There is some evidence that uncontrolled dancing due to drugs has led to heatstroke (McNeill and Parsons, 1996) as well as misuse of drugs in sport (e.g., cycling).

Treatment for heatstroke is immediate and rapid cooling by immersion in chilled water with massage or by wrapping in wet sheet with vigoros fanning with cool dry air. Avoid overcooling and treat shock if present. Suggestions for prevention of heatstroke are medical screening of workers, selection based on health and physical fitness, acclimatization for 5–7 days by graded work and heat exposure, and monitoring workers during sustained work in severe heat. There is some debate over the use of chilled water as cold shock, and rapid rise in blood pressure may cause problems, particularly in the elderly. The use of a warm water bath cooled slowly (Parsons, 2014) or warm water sprayed on a person (lying on a net bed) with fans to aid evaporation will reduce vasoconstriction and thermal shock. The important objective however is to respond quickly and reduce internal body temperature.

NIOSH (2016) notes that the definition of heatstroke has changed in recent years and that two types are now identified. These are the classic heatstroke as described previously and exertional heatstroke which is associated with activity but not a decrease in sweating. Exertional heatstroke occurs, with an associated loss of musculoskeletal function (rhabdomyolysis) and strongly colored urine, among otherwise healthy people in places of strenuous activity, such as in workplaces and at sports events.

Heat Syncope

The clinical feature of heat syncope is fainting while standing erect and immobile in heat. The main predisposing factor is lack of acclimatization, and the underlying physiological disturbance is pooling of blood in dilated vessels of skin and lower parts of the body.

The treatment is to remove the patient to a cooler area and rest in a recumbent position. Recovery should be prompt and complete. Heat syncope should be prevented by acclimatization and, if possible, clothing design to allow evaporation of sweat and allowing intermittent activity to assist venous return to the heart.

Water and/or Salt Depletion Heat Exhaustion

Clinical features of heat exhaustion due to water and/or salt depletion are fatigue, nausea, headache, and giddiness. The skin is clammy and moist, complexion pale, and there is a muddy or hectic flush. The patient may faint on standing with rapid thready pulse and low blood pressure. The oral temperature may be normal or low but the rectal temperature is usually elevated (37.5°C–38.5°C) (99.5°F–101.3°F). If there has been water restriction, urine will be of small volume and highly concentrated (strong color). If there has been salt restriction, urine will be less concentrated and chlorides will be less than $3\,g\,L^{-1}$.

Predisposing factors are sustained exertion in heat, lack of acclimatization, and failure to replace water lost in sweat. Underlying physiological disturbances include dehydration from deficiency of water, depletion of circulating blood volume, and circulatory strain from competing demands for blood flow to the skin and to active muscles.

The treatment is to remove the casualty to a cooler environment, rest them in a recumbent position, and administer fluids by mouth. Keep the patient at rest until urine volume indicates that water balances have been restored. Prevention can be achieved by acclimatization of workers using a breaking-in schedule for 5–7 days with supplement dietary salt (only) during acclimatization. Ample drinking water should be palatable, convenient, and available at all times and to be taken frequently but not excessively during the day.

Water and/or Salt Depletion Heat Cramps

Clinical features of heat cramps caused by water and/or salt are painful spasms of muscles used during work (arms, legs, or abdominal), onset during or after work hours. Predisposing factors include heavy sweating during hot work and drinking large volumes of water without replacing salt loss. Underlying physiological disturbances include loss of body salt in sweat, water intake diluting electrolytes, and water entering muscles causing spasm. Treatment is to provide salted liquids by mouth or more prompt relief by IV infusion. Prevention can be achieved by adequate salt intake with meals particularly in unacclimatized workers. Heat cramps are rarely found in acclimatized people. Hyponatremia occurs when drinking excess amount of water and can be fatal. At particular risk are long distance "fun runners," who misinterpret advice to drink before and during a race.

Prickly Heat

Clinical features of prickly heat are profuse tiny raised red vesicles (blister-like) on affected areas causing pricking sensations during heat exposure. Predisposing factors are unrelieved exposure to humid heat with skin continuously wet with unevaporated sweat. The underlying physiological disturbance is plugging of sweat gland ducts with retention of sweat and inflammatory reaction. Treatment is to apply mild drying lotions ensuring skin cleanliness to prevent infection and prevention is to provide cool sleeping quarters to allow skin to dry between heat exposures.

Anhidrotic Heat Exhaustion

Clinical features of anhidrotic heat exhaustion are extensive areas of skin which do not sweat on heat exposure. They present gooseflesh appearance, which subsides with cool environments and is associated with incapacitation in heat. Predisposing factors are weeks or months of constant exposure to climatic heat with previous history of extensive heat rash and sunburn, and the underlying physiological disturbances are skin trauma (heat rash, sunburn). This causes sweat retention deep in the skin and reduced evaporative cooling causing heat intolerance.

There is no effective treatment for anhidrotic areas of skin. Recovery of sweating occurs gradually on return to a cooler climate. Prevention is to treat the heat rash and avoid further skin trauma by sunburn and providing periodic relief from sustained heat. Mild sunburn, which provides a dead layer of skin above sweat glands, can cause pooling and restrict evaporation.

Transient Heat Fatigue

Clinical features of transient heat fatigue are impaired performance of skilled sensorimotor, mental, or vigilance tasks in the heat. Performance decrement is greater in unacclimatized and unskilled workers. Underlying physiological disturbances are discomfort and physiological strain. There is no treatment although avoidance of boredom and social, organizational, and motivational factors will be important and can contribute to reduction and avoidance along with the acclimatization and training for work in the heat.

Chronic Heat Fatigue

Clinical features of chronic heat fatigue are reduced performance capacity, lowering of self-imposed standards of social behavior (e.g., alcoholic overindulgence), and inability to concentrate. Workers from temperate climates, who arrive for long residence in tropical latitudes and with no effective social and organizational integration, are at risk. Psychological stresses are most important with little evidence of physiological disturbance. Medical counseling and psychological treatment are required for serious cases, with expectation of speedy relief of symptoms on returning home. Orientation on life in hot regions (customs, climate, living conditions, and so forth) will be important for prevention of symptoms. Avoidance of boredom and social, organizational, and motivational factors will be important.

PREVENTION OF HEAT ILLNESS

How to prevent specific heat illnesses is presented in the previous section; however, unless this advice is implemented into a management plan, it may be ineffective. This is important for all systems from sport and military exercises to leisure activities and industrial work. It is particularly important for public health systems. Some systems are presented in the following. It is important to begin with an optimistic (and realistic) approach and to identify that much is known concerning how to avoid heat casualties and that any plan should have the aim of zero casualties. It should also be recognized that ensuring that water and power are available (as public facilities) is essential, as they will be vital resources for cooling the population.

The principles of ensuring sufficient heat loss from the body when exposed to heat (convection, radiation, and evaporation) and manipulation of the relevant variables to create acceptable environments (air temperature, radiant temperature, humidity, air velocity, clothing, and activity) provide the basis for selecting and understanding practical advice for avoiding heat casualties. It is in that context that the advice provided in the following should be considered. This is particularly important as action plans, advice, and guidance have become so extensive that the long lists of actions can cause confusion and make it difficult to identify priorities and make specific plans.

HEAT WAVE ACTION PLANS

The consequences of heat stress and hot weather, including heat waves, are well known and much consideration and research have been coordinated nationally and internationally. The World Health Organization has taken a lead and produced much advice and documentation (Matthies et al., 2008). The knowledge provided so far, in this chapter, has demonstrated that sufficient is known about the causes of heat casualties to determine how to avoid them. It is recognized, however, that this knowledge has to be converted into practical advice and that advice has to be made available in a timely manner to those who can implement it. Heat stress management systems have therefore been developed and implemented. In England, an updated heat wave plan is published every year (Public Health England). "The heat wave plan depends on having well co-ordinated plans in place for how to deal with severe hot weather before it strikes. It builds on our own experience in England and on expert advice from the WHO and the EuroHEAT project (Section 4 of companion document 'Making the case') in developing other national heat wave plans" (PHE, 2018).

The core elements of the plan are strategic planning, alert systems, heat wave preparedness, communicating with the public, working with service providers, working with the community, and monitoring and evaluation. Five alert levels are defined with progressive action. Level 0 is all-year long-term planning; level 1 is heat wave and summer preparedness for the period from 1 June to 15 September; level 2 is when a heat wave is forecast in the next 2–3 days; level 3 is the heat wave action; and level 4 is the declaration of a major incidence with an emergency response required. A heat wave is declared if maximum temperatures are reached for day and night. These range from lower thresholds in the north of England (28°C day and 15°C night) to the South and London at 32°C day and 18°C at night.

The Public Health England and the World Health Organization publications give similar advice for the prevention of heat casualties. This is provided in the following section with some additions and comments (PHE, 2018; Mattheis et al, 2008; WHO, 2012).

ADVICE FOR THE PUBLIC

For the public, advice includes closing windows and shutters during the day (especially those facing the sun) and, if safe to do so, opening windows and shutters at night, when the outside temperature is lower. If the residence is air conditioned, close the doors and windows. Electric fans may provide relief, but when the temperature is above 35°C, fans may not prevent heat-related illness. It is important to drink fluids. Move to the coolest room in the home, especially at night. If it is not possible to keep your home cool, spend 2–3 h of the day in a cool place (e.g., air-conditioned public building, shopping mall, or supermarket). Avoid going outside during the hottest time of the day. Avoid strenuous physical activity. Stay in the shade. Do not leave children or animals in a parked vehicle. Take cool showers or baths. Alternatives include cold packs and wraps, towels, sponging, and foot baths. Wear light, loose-fitting clothes of natural materials. If you go outside wear a wide brimmed hat or cap and sunglasses. Drink regularly and avoid beverages with sugar or alcohol. If anyone you know is at risk, help them to get advice and support. Elderly or sick people living alone should be visited at least daily. If the person is taking medication, check with the treating doctor how they can influence their thermoregulation and their fluid balance.

Keep medicines below 25°C or in the fridge (read the storage instructions on the packaging) and seek medical advice if you are suffering from a chronic medical condition or taking multiple medications. Try to get help if you feel dizzy, weak, anxious, or have intense thirst and headache. Move to a cool place as soon as possible and measure your body temperature. Drink some water or fruit juice to rehydrate. Rest immediately in a cool place if you have painful muscular spasms, most often in the legs, arms, or abdomen, in many cases after sustained exercise during very hot weather and drink oral rehydration solutions containing electrolytes. If children exhibit abnormal behavior (sometimes hyperactivity) then they should sit in a cool bath or play in the shade with cool water, while calm and stationary. If problems continue, consult a medical doctor.

Medical attention is needed if heat cramps are sustained for more than 1 h; consult your medical doctor if you feel unusual symptoms or if symptoms persist. If one of your family members or people you assist presents hot dry skin and delirium, convulsions, and/or unconsciousness call the doctor/ambulance immediately. While waiting for the doctor or ambulance move him/her to a cool place and put him/her in a horizontal position and elevate legs and hips, remove clothing, and initiate external cooling, such as with cold packs on the neck, axillae, and groin; continuous fanning; and spraying the skin with water at 25°C–30°C. Measure the body temperature. Do not give acetylsalicylic acid or paracetamol. Position unconscious person on their side.

ADVICE FOR SERVICE PROVIDERS

For service providers, details of helplines, social services, ambulances, cool spaces, and transport need to be provided on any information material. Provide access to cool spaces and ensure active assistance for those most at risk. In addition to the general information, information for the elderly (and the very elderly) and people with chronic diseases should contain: practical tips (e.g., for keeping cool and well hydrated), first aid treatment, and important contact details for social and medical services as well as the ambulance. Other population groups that may need to be considered for specific information may include workers, athletes, tourists, and parents of infants. Particularly for the elderly, the socially isolated, and the homeless, passive information through leaflets and brochures has proven not to be sufficiently effective, and other more active approaches need to accompany any public health measures, such as a buddy system, visits, and phone calls. The use of social media systems has great potential and the use of existing social media advice can be helpful but will vary in validity and quality.

Vulnerable population groups include those with diabetes mellitus, other endocrine disorders, organic mental disorders, dementia, Alzheimer's, mental and behavioral disorders due to psychoactive substance use, alcoholism, schizophrenia, schizotypal and delusional disorders, extrapyramidal and movement disorders (e.g., Parkinson's disease), cardiovascular disease, hypertension, coronary artery disease, and heart conduction disorders.

They also include those with diseases of the respiratory system including chronic lower respiratory disease (COPD, bronchitis) and those with diseases of the renal system including renal failure, and kidney stones. Acute diseases include infections, fever, gastroenteritis, and skin infections are also risk factors for heat-related mortality (Kilbourne, 1997).

ADVICE FOR GENERAL MEDICAL PRACTITIONERS

General medical practitioners are encouraged to develop a proactive approach by the following:

- Understanding the thermoregulatory and hemodynamic responses to excessive heat exposure
- Understanding the mechanisms of heat illnesses, their clinical manifestations, diagnosis, and treatment
- Recognizing early signs of heatstroke, which is a medical emergency
- Initiating proper cooling and resuscitative measures (for early signs and out-of-hospital treatment refer to the separate information sheet on treatment of heatstroke and other mild heat-related illnesses)
- Being aware of the risk and protective factors in heat wave-related illness
- Identifying the patients at risk and encouraging proper education regarding heat illnesses and their prevention (education of guardians of the old and infirm and infants is important)
- Including a pre-summer medical assessment and advice relevant to heat into routine care for people with chronic disease (reduction of heat exposure, fluid intake, medication)
- Being aware of the potential side effects of the medicines prescribed and adjusting dose if necessary, during hot weather and heat waves

- Making decisions on an individual basis since there are—according to current knowledge—no standards or formal advice for alteration in medications during hot weather
- Being aware that high temperatures can adversely affect the efficacy of drugs, as most manufactured drugs are licensed for storage at temperatures up to 25°C; ensuring that emergency drugs are stored and transported at proper temperature
- Being prepared to monitor drug therapy and fluid intake, especially in the old and infirm and those with advanced cardiac diseases

Note that a drinking regime is required as any fluid lost due to sweating and other mechanisms in the heat (e.g., related diarrhea) will need to be replaced so that dehydration can be avoided.

Educate, counsel, and inform patients regarding the following:

- The importance of adhering to the recommendations spelt out in the leaflet for the public
- Individual adjustments of behavior, medication, and fluid intake according to the clinical status
- Contact details of social and medical services, helplines, and emergency services

ADVICE FOR RETIREMENT AND CARE HOME MANAGERS

PHE (2018) provides a specific information sheet (recommendations for the public during heat waves) for advice on how to keep facilities cool and ensure that patients and residents keep out of the heat, cool, and hydrated. Mangers should monitor indoor temperatures, provide at least one cool room (e.g., air-conditioned room below 25°C), and move residents to this cool area for several hours each day. Ask general practitioners to review clinical management of residents at risk, for example, due to chronic disease. Monitor fluid intake. Offer nonalcoholic, unsweetened beverages. Monitor body temperature, pulse rate, blood pressure, and hydration. Monitor closely for any early signs of heat illness and initiate appropriate treatment. Inform and train staff and increase staffing levels if necessary.

WATER, FANS, AND AIR CONDITIONING

Although there may be exceptions and complications, it is generally true that no one dies of heat stress in cool water or in an air-conditioned room at low enough temperature. Immersion in a cool bath is a powerful method of avoiding heat stress. Even immersing hands, arms, legs, or feet in cool water provides great relief and can also be performed where clothing cannot be removed (e.g., protective clothing). Fans require consideration as advice is sometimes not to use them at high temperatures (>35°C). This is not good advice from a heat transfer perspective, as evaporation will more than compensate for any gain by convection (the Lewis Relation). I experienced this when acclimating international golfers in a thermal chamber before they went off to a hot climate to play in a tournament. The chamber was at 40°C air temperature with full fans. On entry, the team objected to the fans, but once they started sweating, they quickly recognized their value. There are problems, however. It is useful to spray water on the face, hair, or other parts of the body as well as providing fans. However, moving air dries the mucus in the airways and care should be taken not to initiate respiratory problems. Moving air across the body is very helpful, but care must be taken not to cause other problems. Survivors of heat waves in cities are often those who have air conditioning. Advice should be given to ensure minimal clothing and low activity to reduce the requirements for low air temperatures. Maintaining the temperature at 25°C in residences and offices should avoid discomfort due to heat if formal protocols, such as strict adherence to uniforms and no moving around, are relaxed. Unnecessary reductions in temperature will require great electrical power, across a town or city, and may cause major problems as a surge in electricity supply may cause power cuts.

DRUGS

Drugs can directly affect thermoregulation including cardiac output, heart rate, blood circulation, and sweating. Thermoregulation may also affect the influence of drugs. Dehydration caused by exposure to the heat may affect the efficacy of drugs, and the drugs may cause dehydration which will be a risk factor in heat stress. Drugs with anticholinergic effects are potent inhibitors of sweating. Antipsychotic drugs may in addition interfere with the central control of the body temperature. Vasodilators including nitrates and calcium channel blockers can worsen hypotension in vulnerable patients. Toxicity of drugs with a narrow therapeutic index, such as digoxin or lithium, may be enhanced.

WORKING PRACTICES FOR HOT ENVIRONMENTS

Working practices to ensure acceptable thermal strain when subject to heat are a general consideration irrespective of context. However, they are usually considered in terms of the working environment. Kjellstrom et al. (2013) point out the problems at work that may occur due to future global warming. Malchaire et al. (1999) and Bethea and Parsons (2002) provide risk assessment methodologies for work in the heat. The National Institute of Occupational Safety and Health (NIOSH) in the United States (NIOSH, 1986, 2016) provide working practices for hot environments in terms of engineering controls, work and hygiene, administrative controls, heat alert programs, body cooling and protective clothing, and performance degradation. Engineering controls suggest moving away from the heat source if possible but also increasing air velocity, reducing radiation and the temperature of surrounding surfaces and increase evaporation using moving air, reduced humidity, and the evaporation of water from the person with good ventilation. Hygiene and administration systems include redesigning work, limiting exposure times, reducing metabolic load, acclimatization, and health and safety training. For a heat alert program, preparation must be made with a plan and training when the heat arrives. Protective clothing causes particular risk, and active cooling should be used if possible and careful monitoring will be required. When wearing protective clothing, performance degradation may be expected, and this will be augmented by the heat.

ISO 12894 (2001) provides screening methods for workers to ensure that those exposed are fit for work. ISO 15265 (2004) provides a risk assessment methodology, and ISO TR 16595 (2018) is a proposed new standard under development for describing working practices for hot environments.

COOLING OF PEOPLE INVOLVED IN SPORT

Racinais et al. (2015) in their consensus document on recommendations for conducting sport in the heat, suggest that precooling (e.g., cool drinks and ice slurries) may have an advantage in sports involving sustained exercise. On course, cool showers and fans during long distance running will be useful to cool runners. When a runner drinks a small amount of water from a bottle during a race, instead of throwing the rest away, which seems common practice, pouring the rest over their head will be useful for cooling. For post-cooling, drinks and ice slurries will be useful as well as various garments, such as ice vests and the use of wet towels. Fans and water immersion, if available, are also effective methods of cooling. Care must be taken to ensure that actions are appropriate and do not inhibit thermoregulatory cooling due to sweating or misinterpret the need to keep warm as, when stopped, a very high level of metabolic rate will reduce to resting levels but sweating will continue. A thermal audit is required, with experimental evaluation, for each sport, to ensure that guidelines are appropriate. Each sport must be considered in terms of its characteristics. In tennis, for example, why not use a glass air-conditioned booth with fans, between sets, instead of players sitting in the heat (albeit with a boy or girl holding a sunshade). Quarterly breaks and half time can be used to great effect if used appropriately. Experimental research and practical evaluation will develop methods for each sport with the use of measurements of heat strain and thermal audits described in previous chapters.

11 Human Performance and Productivity in the Heat

HUMAN PERFORMANCE MODEL

In his editorial for the journal "*Industrial Health*," Parsons (2018) described an international initiative to standardize the way we consider the effects of the environment on human performance. Three effects are described as follows: those on the capacity to carry out tasks, time off task due to distraction, and time off work for reasons of safety.

Human performance has meaning only in terms of carrying out a task and implies that there is a "level" of performance that can be represented on a scale determined by a dependent variable (time, accuracy, achieved or not, user satisfaction, and so forth). Productivity is usually an organizational concept often related to economic success, but it can also be related to individual output, such as the number of patients recovered in a clinic or the number of graduates from a university. In management terms, it will be related to organizational mission, strategy, tactics, and business plan. An essential starting point for describing the effects of heat stress on human performance and productivity is therefore to set up a model to allow us to consider the problem, including definitions of performance and productivity and how we will measure them.

The draft definitions provided for the proposed new international standard (Technical Report) are as follows: human performance is the extent to which a person carries out a task or a combination of tasks; productivity is the amount a person, group, or organization produces (ISO NP TR 23454-1, 2018).

THE EFFECTS OF HEAT STRESS ON THE CAPACITY TO CARRY OUT TASKS

The thermoregulatory response of the body provides an indication of how heat may interfere with human performance at carrying out tasks. As the tissues become warm, they will increase in metabolism and blood flow will provide effective transport of resources to, and waste products away from, cells. It may be expected that performance at tasks involving movement might improve as would the performance of muscles. Sweating might be expected to interfere with tasks involving fine manual dexterity and grip. A practical effect may be that glasses or safety goggles may mist up due to evaporated sweat or condensation (NIOSH, 2016) or clothing, especially impermeable gloves, may fill with sweat and impair finger dexterity. In an experiment in a thermal chamber, investigating heat stress while wearing a full impermeable suit, I observed fingers of gloves extending by over 10 cm due to sweat accumulation.

There have been numerous studies into the effects of heat stress on human task performance, and effects have been found but they are not easy to predict. Knowing the level of heat stress does not allow us to predict task performance with accuracy. That is probably due to inconsistency in repeated trials, individual differences, individual disposition (e.g., motivation and all sorts of personal factors), and more. Results have, however, provided sufficient information for useful general guidance to be provided.

The international standards organization has made an initial proposal to define a human performance model as consisting of cognitive, physical, and perceptual motor tasks (ISO NP TR 23454-1, 2018). In practice "in the real world" tasks are complex and involve combinations of human activities. Studies that look at, a range of, individual components of complex (multicomponent) tasks or jobs will provide an indication of performance at overall tasks by integrating results. For a more definitive indication, studies can be conducted on "actual" tasks of interest, but they may be inconvenient or expensive to carry out and results will have important specific, but limited general, application. Cognitive tasks include those on memory, logical reasoning, learning, signal detection, vigilance, information processing, and decision-making. Physical tasks (manual performance) include fine dexterity, gross motor performance, lifting and handling, and endurance. Perceptual motor tasks include tracking where a person detects and follows an object and manually tracks it with a controller, or controlling a robot or remotely piloted vehicle such as a drone.

Hancock (1982) emphasizes the importance of task categorization when considering the limits of human performance in extreme heat. He concludes that heat stress can cause a decrement in three categories of task. These are mental and cognitive skills, tracking, and performance at dual tasks. In a comprehensive review, he identifies 85°F (29.4°C) as the Effective Temperature (ET) above which decrements in performance may be expected. It may be noted that Pepler (1964) provides the same value for (air) temperature which he converts to 81°F (27.2°C) ET. Hancock (1982) notes that for the conditions investigated in the studies, this is close to the upper limit of the prescriptive zone or when heat stress approaches noncompensable levels and internal body temperature begins to rise. He notes that the decrements in performance at a range of tasks and over task categories correlate well with internal body temperature and that the higher the level of skill, the less influence of the heat stress on performance. This was clearly shown by Mackworth (1950) in Morse code operators and is explained in terms of workload by Parsons (2014).

Pepler (1964) provides a comprehensive review of the extensive number of studies on the effects of heat on human performance up to 1964. He also notes that level of skill, acclimatization, and motivation are important factors. Early studies of productivity in industry had demonstrated that as conditions varied from those that created comfort, productivity fell and accident rates went up (Vernon, 1919a,b, 1920). Investigations into long-term effects of heat (e.g., so-called tropical fatigue), when moving to and living in a hot climate, were inconclusive and confounded with other factors. Meese et al. (1984) transported a climatic chamber to factory "car parks" in South Africa and conducted tests on the effects of moderate heat on actual factory workers from the shop floor. Using a battery of tests to measure a range of attributes, they found little evidence of effects on cognitive performance but some improvement in physical performance as conditions became hotter but then a fall in performance as they became more extreme. However, in an applied study in an automobile parts manufacturing plant, Ciuha et al. (2017) noted that the management reported a 13% drop in productivity during a heat wave.

Ramsey and Kwon (1988) note the work of Wing (1965) who provided a tentative relationship between ET [later converted to wet-bulb globe temperature (WBGT)] values and when a fall in mental performance could be expected for sedentary jobs. An exponential decay was proposed with a rapid decay in WBGT limit values from 45°C WBGT for around 10 min exposure down to around 35°C WBGT for 1 h of exposure then a slower drop to 4 h to around 30°C WBGT. This was not supported by a more comprehensive analysis by Ramsey and Kwon (1988) who reviewed over 150 studies and found little evidence of significant effects of heat stress on mental performance. They found some effects on perceptual motor performance, and for both of those categories, it could be argued that if a limit value had to be chosen, then 30°C WBGT may be a reasonable starting point for people wearing normal clothing and conducting light activity. For the effects of heat stress on manual performance, improvements may be expected, due to warming of muscles and increased dexterity and speed of movement, but that as physiological strain increases (>33°C WBGT) performance may fall. Ramsey et al. (1983) considered unsafe working behaviors as a dependent variable and provided WBGT limit values which increased from a minimum at 22°C WBGT to a high proportion of unsafe behaviors at 35°C WBGT, the upper limit for which there were data. In the absence of conclusive information on the effects of heat stress on human task performance, it would seem reasonable to conclude that existing limits for physiological responses (e.g., ISO 7243, 2017) could be used as a warning for possible loss of performance but that it is not certain that it will occur. This would then include clothing and activity levels as well as acclimatization. However, motivation and skill level would still remain as unaccounted for influential parameters. A high level of skill and motivation may overcome any negative effects of heat stress; however, a high level of motivation may encourage a person to carry on, or increase exposure to, the heat stress, as they do not recognize, or deny, a loss in capacity.

We can speculate how we could predict how heat might affect task performance. It seems reasonable from the studies of Ramsey and Kwon (1988) that there will be no loss of performance below WBGT limits as specified in ISO 7243 (2017). For a lightly clothed sedentary person, this would be around 30°C WBGT. Sweating required (Sw_{req}) and achieved (Sw_p) will indicate strain for hot environments within the prescriptive zone and can be related to task performance. Internal body temperature could be used as an indicator of likely loss in performance above the upper limit of the prescriptive zone until collapse. Motivation will be important. We could speculate up to 20% loss in task performance [when compared with optimum (e.g., comfort)] up to the upper limit of the prescriptive zone for a motivated person. If the person is not motivated at all, then performance will be less. For a highly motivated person, performance decrement will reduce (ignoring any effects of over arousal).

Above the upper limit of the prescriptive zone, performance may fall exponentially from 80% to 0%. So based upon limits set for WBGT (ISO 7243, 2017) and using the body heat equation (ISO 7933, 2004) or a thermal model (Gagge et al., 1971), and taking account of motivation, and maybe type of task, we could estimate how much heat could affect task performance. That is the capacity to carry out a task in the heat but not taking account of any distraction. Evidence is not conclusive, however, and it may be more valid to conclude that as long as heat does not cause heat illness or collapse, there will be no significant effect of heat on the capacity of a person to carry out a task. Any predictable significant effects will be due to distraction and time off a job for reasons of safety.

TIME OFF TASK DUE TO DISTRACTION

When heat stress causes discomfort, dissatisfaction, intolerance, frustration, apathy, and so on, a person will be distracted from performing a task and take time off to attend to the heat possibly in an attempt to do something about it. Adjusting or taking off clothing, taking a break, opening a window, attempting to work out how to use the air conditioning system, wiping sweat, and more, will all be examples of how task performance can be reduced to zero for a period of time. Parsons (2014) defined thermal distraction as a tendency of a person to attend to a thermal state (e.g., hot and sweaty) instead of performing a task. Social interaction may add to distraction as a group may confer and discuss how to alleviate any heat stress.

That heat can cause distraction when performing a task is of little doubt, but when it will occur and to what extend will depend upon the level of heat stress and the disposition of the person in the heat. There have been few studies of the relationship between how much heat will cause how much distraction. There have been a large number of studies into how thermal environments determine how people feel (Rohles and Nevins, 1971; Fanger, 1970; Gagge et al., 1973; Webb and Parsons, 1997). A thermal index can be used to predict how people feel from the six basic parameters that make up heat stress (air temperature, radiant temperature, air velocity, humidity, clothing, and activity). These include the predicted mean vote (PMV, Fanger, 1970; ISO 7730, 2005), the standard effective temperature (Gagge et al., 1973), and the universal thermal climate index (Jendritzky et al., 2012).

If we consider dissatisfaction with the thermal environment as a possible stimulus for distraction, then the predicted percentage of dissatisfied (PPD) index, derived from the PMV value, is probably as valid as others and provides a useful index for considering degree of distraction. Although the PPD is intended to predict the percentage of people in a large group as being dissatisfied, we can use it here as a degree of dissatisfaction. It has the advantage of including all six basic parameters of air temperature, radiant temperature, air velocity, humidity, clothing, and activity, and that it has been accepted in ISO 7730 (2005).

If distraction is related to dissatisfaction, then it will start from near zero distraction (actually 5% is the minimum value for the PPD) at comfort and increase to 100% when hot. Motivation will determine how much dissatisfaction will cause how much distraction. For a person with no motivation to perform the task or job, distraction could be 100% and performance would be zero. A motivated person may only gradually be distracted as dissatisfaction rises, and a highly motivated person may not be distracted at all as any dissatisfaction will be ignored in favor of achieving a level of performance.

TIME OFF JOB FOR SAFETY REASONS

Limits to ensure that people can maintain health and safety are provided in terms of the WBGT index in ISO 7243 (2017) and the Predicted Heat Strain method described in ISO 7933 (2004). If heat stress or physiological limits are exceeded then there will be likely to be unacceptable thermal strain (based upon empirical evidence, predicted internal body temperature, dehydration, and so on). Beyond the limits, exposure to the heat will be considered unsafe. If conditions are unacceptable, then exposure time must be restricted. An analytical calculation is conducted to determine rest pauses to ensure that heat stress exposures are within limits. ISO 7243 (2017) uses time-weighted averaging of exposures (e.g., to work and rest areas) to determine rest time. If we ignore the possibility of redesign of work (e.g., working in cooler areas), then in terms of human performance and productivity, rest time will be time off work and hence zero performance. It should be noted that motivation will be important at high levels of heat stress, but that the safety limit values will not allow it to become influential.

PUTTING IT ALL TOGETHER

The aforementioned discussion has rationale and some validity. More research is required to determine the effects of heat on the capacity to carry out tasks, cause distraction, and to predict time off work. Much work has been done, however, and it maybe that there is no accurate definitive answer. In such circumstances, it is often useful to provide a best estimate and at least provide a standardized method that can be used across the world. Actually, this is why the ISO standards activity proposes a technical report rather than a full international standard. It allows experience with a method for later decision about whether it would be useful as a standard. It maybe that the best any method could provide is a confirmation that it is unlikely that there would be an effect on performance below a certain level of heat stress and an increasing possibility of loss in performance and productivity above that level.

A final speculation is how to bring the three effects together to make an assessment in practice. A first stage is to measure the heat stress. That is, to quantify the heat stress in terms of the WBGT value or other index including an analytical index, such as the PHS which would involve measurement of air temperature, radiant temperature, air velocity, humidity, clothing, and activity. The total time for work would be modified by the time allowed for safety. Rest time can be assume to have zero performance or productivity and is independent of motivation. Work time would be the only time where other affects (distraction and capacity to perform a task) could occur. Distraction can be related to dissatisfaction and modified for motivation. Capacity to perform tasks will have a threshold below which there will be no effect and above which will have an increasing affect depending on the task and motivation. Overall, it would seem that total performance when compared with optimum conditions would be estimated by multiplying the time at work (taking account of any rest periods for health and safety reasons) by the time on the task (taking account of time off task due to distraction) by the capacity to perform the task in the heat (taking account of motivation). In a hot environment for an 8-h shift, where work and rest are 50%, distraction is 40% (so 60% on task) and capacity to carry out the work task is 90%, we would estimate a performance of 50% × 60% × 90% = 27% productivity or a loss of 73% and hence almost 6 h in a working day.

An example of the process can be seen for a comfortable condition and for a hot condition. Suppose the thermal environment was comfortable, the total time available for work (e.g., 8 h including breaks) would not be reduced for health and safety reasons of heat stress (i.e., no reduction so 100% possible performance) whether motivation was high, moderate, or low.

Minimum loss in performance due to distraction because of heat can be expected in comfortable conditions, even for a person with low motivation. Dissatisfaction would be predicted by the PPD index at 5%, so that could be a starting loss in performance to include any individual differences (some people will be warm when the average of a group is comfortable). If we select a value for high motivation as $m = 1$; moderate motivation as $m = 0.5$, and low motivation as $m = 0$, then multiplying the PPD value by the level of motivation [actually $(1 - m)$] provides a value related to the loss in performance due to distraction due to heat. So for a highly motivated ($m = 1$) person in comfort, we can predict (1 – 1) loss due to motivation × 5% dissatisfied = 0 distraction. For a moderately motivated person, we can predict (1 – 0.5) motivation × 5% dissatisfied = 2.5% distraction. For a person with low motivation ($m = 0$), we can predict (1 – 0) due to motivation × 5% = 5% loss in performance due to distraction. Time on the task taking account of distraction is therefore 100% if highly motivated, 97.5% for moderate motivation, and 95% for low motivation.

The final calculation is to consider the capacity to carry out the task. As we are considering comfort conditions, performance will not be reduced. So, combining the effects of health and safety, distraction, and loss in capacity together, the total performance in a comfortable environment is no loss in performance and productivity for a highly motivated person or group, 2.5% drop in performance for a moderately motivated person or group, and a 5% drop in performance and productivity for a person or group with low motivation. That is compared to a highly motivated group in comfortable thermal conditions.

Consider the above "model" for hot conditions. Suppose that we measure the heat stress in terms of WBGT values as 35°C in normal work clothing conducting light activity. Based upon a weighted average of work and rest conditions, for an 8-h day (with normal breaks and ignoring fatigue), we find that 30 min rest is required every hour. That is a 50% reduction (from a total of 8 to 4 h of work) in performance and productivity. This will not depend upon the motivation level of the person.

Performance loss due to distraction will be related to dissatisfaction and the PPD value which in these hot conditions will be 100% (although a less comfort orientated index may be needed). If, as before, we select a value for high motivation as $m = 1$, moderate motivation as $m = 0.5$, and low motivation as $m = 0$, then multiplying the PPD value by $(1 - m)$, the level of motivation provides a value related to the loss in performance due to distraction due to heat. So for a highly motivated person in comfort, we can predict (1 – 1) loss due to motivation × 100% dissatisfied = 0 distraction. For a moderately motivated person, we can predict (1–0.5) motivation × 100% dissatisfied = 50% distraction. For a person with low motivation, we can predict (1 – 0) due to motivation × 100% dissatisfied = 100% loss in performance due to distraction. So, for a highly motivated person, productivity and performance will be for the full 4 h of work. For a moderately motivated person, it will be for 2 of the 4 h as they will be distracted for the other 2 h, and for a person with low motivation, there will be zero performance and productivity as they will be continuously distracted.

The final consideration is the capacity to perform the task. Suppose the capacity to perform a task in the heat will fall to 90% of performance at comfort levels. We again select a value for high motivation as $m = 1$, moderate motivation as $m = 0.5$, and low motivation as $m = 0$. For a person or group of high motivation, the capacity to carry out the task will be $m = 1$ for motivation × 90% for capacity to carry out the task = 90%. So, performance will be 90%. For a moderately motivated person or group, it will be 0.5 for motivation × 90% capacity to carry out the task = 45% performance, and for people who have low motivation, it will be 0 for motivation × 90% for capacity to carry out the task = 0% performance.

For the hot conditions, therefore, performance is for a highly motivated person 50% for safety × 100% as time on task when not distracted × 90% as the performance possible when carrying out the task in the heat = 45% of the performance that would be possible if the conditions were comfortable. In terms of an 8-h work period, that would be 4 h of rest and 4 h of nondistracted work at 90% of optimum performance. A similar calculation can be carried out for other levels of motivation.

In fact, the evidence for loss of capacity to perform cognitive, manual, or perceptual motor tasks in the heat is not clear until collapse occurs. For hot conditions, therefore, distraction and time off job will be the main factors. It may be reasonable to conclude that the use of existing international standards for assessing heat stress, will therefore provide not only an indication of safety limits but also an indication of any loss in performance and productivity.

Skin Burns and Contact with Hot Surfaces

12

SAFE SURFACE TEMPERATURES

When human skin touches a hot surface, the skin heats up at a rate and to a level that depends upon the temperature of the hot surface, its material type, and contact time. If the heat transferred increases the temperature of the skin to a level above threshold values, then the cells of the skin will suffer damage and the contact will cause a burn.

To select, design, and specify maximum (touchable) surface temperatures for a machine, in an aircraft cockpit, in a kitchen, for central heating systems, such as radiators, on toys (containing batteries) that heat up, and more, we need to know what temperatures of what surfaces will be unacceptable and cause a burn. We can then provide limit values and avoid people burning themselves. These are safe surface temperatures. As it is unethical to burn humans, it was "done" to young pigs. Moritz and Henriques (1947) exposed young pigs to a range of temperatures produced by flowing hot water (and sometimes oil) over their skin. They then observed whether a burn occurred and noted the severity of the burn (for a fuller description, see Moritz and Henriques, 1947; McIntyre, 1980; Parsons, 2014).

HUMAN SKIN

The skin is a complex active structure containing glands and pores, hairs, nerve endings that respond to high temperature, low temperature, touch, pain and pressure, a range of types of cells, blood, lymph, and more. It is the part of the body which comes into contact with the outside world. It can be turgid, its blood vessels can be vasodilated or vasoconstricted, and it can be wet with sweat (see Montagna and Parakkal, 1974; Wood and Bladon, 1985; Millington and Wilkinson, 2009).

The skin can be considered as three layers, and these are used to define the severity of burns. The outer, thinner layer is the epidermis which has a basement membrane which generates cells that progress towards the surface (the epithelium) such that outer dead cells eventually fall from the body so that the epidermis is continuously renewed. The dermis is a thicker layer below the basement membrane and contains nerve endings, sebaceous glands (for the hair), and is permeated by sweat pores and hair. The third layer is the adipose tissue which contains the sweat glands, the hair base, and the pressure sensors.

BURNS

A burn which damages the epidermis but not the basement membrane is called a superficial burn and is a first-degree burn, although classification systems vary. They are painful at first and heal within a few days. A superficial partial-thickness burn (ISO 13732-1, 2001) extends through the epidermis into the dermis. They cause redness and blisters and take about 3 weeks to heal. It is the surface temperatures that cause this type of burn that are specified as limits for products and physical environments. They are sometimes called first-degree burns, but mostly second-degree burns. They are painful as they do not destroy the pain sensors, and they heal without leaving a scar as they do not destroy the basement membrane. More severe burns destroy the basal membrane and penetrate further into the dermis than superficial partial-thickness burns (second-degree or partial-thickness burn) or are deeper whole thickness burns (third degree) which destroy the whole of the skin and maybe more. The size of the area, location, and percentage of the skin burned also reflect severity and influence categorization of the burn (Figure 12.1).

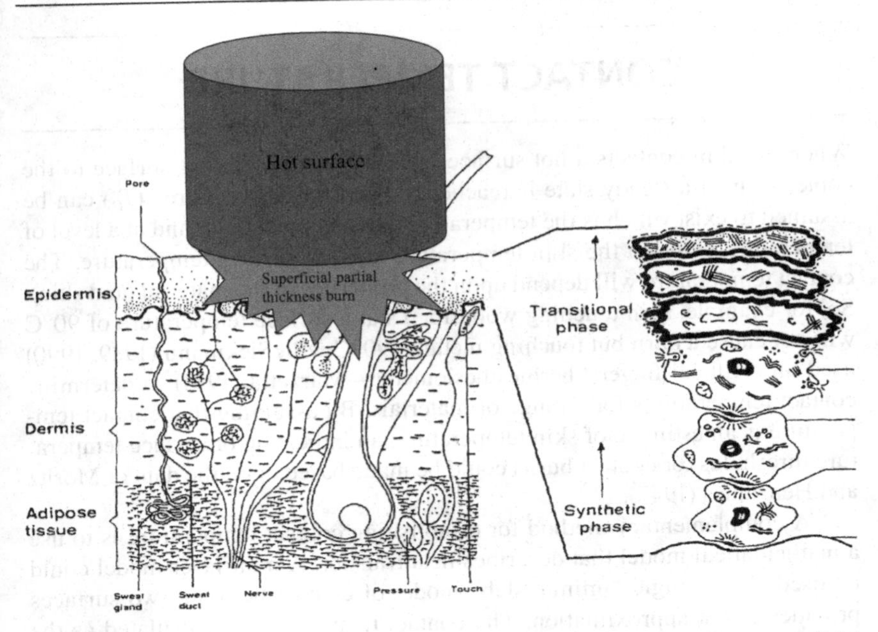

FIGURE 12.1 Diagrammatic representation of a hot surface causing a superficial partial-thickness burn to human skin. (Modified from Parsons, 2014.)

BURN THRESHOLDS

Moritz and Henriques (1947) produced two curves to summarize their results from burning pigs (and sometimes themselves) with flowing water and oil. They provide thresholds of skin temperature with contact time. Skin temperatures above which a burn would be expected are at 70°C for 1 s of contact and reduce exponentially to around 46°C for 6 h of contact. Skin temperatures below which a burn would not be expected are at 65°C for 1 s of contact and reduce exponentially to around 44°C for 6 h of contact. The study also investigated additional factors, such as combinations of exposures with different contact times and pressure on the skin. There are other studies that have investigated when burns will occur (e.g., Sevitt, 1949); however, the work of Moritz and Henriques (1947) is extensive and is often taken as providing the definitive data for specifying when a skin burn will occur (Parsons, 2014).

CONTACT TEMPERATURE

When the skin contacts a hot surface, heat flows from the hot surface to the cooler skin. If a steady state is reached, a "contact temperature" (T_C) can be assumed to exist which is the temperature at the contact point and at a level of temperature between the skin temperature and the surface temperature. The contact temperature will depend upon the properties of the material (and skin). So, for example, skin touching wood for 1 s at a surface temperature of 90°C will not cause a burn but touching metal at 90°C will. Siekmann (1989, 1990) used an artificial finger (thermesthesiometer—Marzetta, 1974) to determine contact temperatures for a range of materials. By assuming that contact temperature is an estimate of skin temperature, an indication of surface temperature thresholds for contact burns could be made based upon the data of Moritz and Henriques (1947).

A complementary method for establishing contact temperature is to use a mathematical model that describes heat transfer. A multilayer model could be used, but a single "infinite slab" model of contact between two surfaces provides a first approximation. The contact temperature is calculated as the weighted average of thermal penetration coefficients (b) of the two materials (b_1 and b_2) where $T_C = (b_1 \cdot t_1 + b_2 \cdot t_2)/(b_1 + b_2)$; t_1 and t_2 are the temperatures of the two surfaces (slabs). The thermal penetration coefficients ($\mathrm{J\,s^{0.5}\,m^{-2}\,K^{-1}}$) is the square root of the multiple of the thermal conductivity (K), specific heat (c), and density (ρ). McIntyre (1980) gives thermal penetration coefficients for the skin as 1,000; for metals as >10,000; glass as 1,400; and wood as 500. For a skin temperature of 33°C and surface temperature of 100°C, the contact temperatures are as follows: for metal, 94°C; glass, 72°C; and wood, 55°C. A 1-s exposure and perfect contact may mean a superficial partial-thickness burn when touching metal, possible burn when touching glass, and no burn when touching wood. An assumption is made that surface temperatures remain constant whereas, for glass, for example, thin glass may decrease in temperature on contact. Parsons (1993, 2003, 2014) provides a comprehensive consideration of the subject and proposes the equivalent contact temperature index (T_C,eqiv) which additionally takes account of the skin condition and nature of contact.

The results of Moritz and Henriques (1947) and those of Siekmann (1989, 1990) were used to produce a European Standard (EN 563, 1994) that provided "safe surface temperature limits" for the surfaces of machines. This was later developed internationally (ISO 13732-1, 2001) to apply to "all" touchable surfaces (not just those of machines) and all populations (not just workers).

ISO 13732-1 (2001) ERGONOMICS OF THE THERMAL ENVIRONMENT—METHODS FOR THE ASSESSMENT OF HUMAN RESPONSES TO CONTACT WITH SURFACES. PART 1: HOT SURFACES

ISO 13732 is a three-part standard. Part 1 published in 2006 is concerned with skin contact with hot surfaces. Part 2 is concerned with how surfaces of "moderate" temperature feel and Part 3 is concerned with skin contact with cold surfaces. Part 1 presents data, graphically, for contact periods from 0.5 to 10s, depicting three areas (1. no burn, 2. burn threshold, 3. burn). These are a curve depicting an area of surface temperatures above which contact is expected to cause a superficial partial-thickness burn (3. burn); an area depicted by a lower curve below which the surface temperature would not cause a superficial partial-thickness burn (1. no burn). The area in between the two limits is an area of uncertainty where a burn may occur, and increasing temperature provides increased risk (2. burn threshold).

Limit values for contact with (solid) hot surfaces are provided for contact times from 0.5 (e.g., unintentional contact so related to reaction time) to 10s, graphically as described previously and for single values (no areas or ranges) between 1min to over 8h. A linear interpolation between curves and single values is required between 10s and 1min of contact. An analysis of the contact event is needed to establish a limit value (see Table 12.1). The standard applies to all hot solid surfaces and all populations. It does not apply to burns of >10% of the skin or on the head or airways.

A burn threshold is defined in ISO 13732-1 (2001) as a surface temperature defining the boundary between no burn and a superficial partial-thickness burn, caused by contact of the skin with this surface for a specified contact period.

Burns are classified into three levels, depending on severity. These are superficial partial-thickness burns where the epidermis is completely destroyed but the hair follicles and sebaceous glands as well as the sweat glands are spared; deep partial-thickness burns where a substantial part of the dermis and all sebaceous glands are destroyed and only the deeper parts of the hair follicles or the sweat glands survive; and whole thickness burns where the full thickness of the skin is destroyed and there are no surviving epithelial elements. Only superficial partial-thickness burns are considered in this standard.

TABLE 12.1 Burn Thresholds. Surface temperatures (°C) and contact times (s, unless specified) that cause superficial partial-thickness burns on contact with human skin.

	CONTACT TIME (S)							
MATERIAL	*0.5*	*1.0*	*2.0*	*4.0*	*10.0*	*1 MIN*	*10 MIN*	*≥8 H*
Smooth, Bare, Uncoated Metal								
Minimum for definite burn	73	69	67	63	59	51	48	43
Maximum for definite no burn	67	64	62	58	54	51	48	43
Smooth Ceramic Glass and Stone								
Minimum for definite burn	90	85	81	76	70	56	48	43
Maximum for definite no burn	84	79	75	72	66	56	48	43
Smooth Plastics								
Minimum for definite burn	99	93	87	82	76	60	48	43
Maximum for definite no burn	91	85	80	76	70	60	48	43
Smooth Wood								
Minimum for definite burn	151	138	127	116	103	60	48	43
Maximum for definite no burn	125	114	104	95	85	60	48	43

A linear interpolation should be taken between 10 s and 1 min.

Bibliography

ACGIH, 2017, Heat stress, TLVs and BEIs: Threshold limit values for chemical substances and physical agents & biological exposure indices, Cincinnati, OH: ACGIH.

ACGIH, 2018, TLVs and BEIs based on the documentation of the threshold limit values for chemical substances and physical agents and biological exposure indices, Cincinnati, OH: ACGIH.

ACSM, 1984, Prevention of thermal injuries during distance running, *Medicine and Science in Sports and Exercise*, 16(5), 427–443.

Ainsworth, B. E., Haskell, W. L., Herrmann, S. D., Meckes, N., Bassett, D. R., Tudor-Locke, C., Greer, J. L., Vezina, J., Whitt-Glover, M. C. and Leon, A. S., 2011, 2011 compendium of physical activities, *Medicine & Science in Sports & Exercise*, 43(8), 1575–1581. doi:10.1249/mss.0b013e31821ece12.

Al-Ajmia, F. F., Loveday, D. L., Bedwell, K. H. and Havenith, G., 2008, Thermal insulation and clothing area factors of typical Arabian gulf clothing ensembles for males and females: Measurements using thermal manikins, *Applied Ergonomics*, 39(3), 407–414.

Asamoah, B., Kjellstrom, T. and Östergren, O., 2018, Is ambient heat exposure levels associated with miscarriage or stillbirths in hot regions? A cross-sectional study using survey data from the Ghana Maternal Health Survey 2007, *International Journal of Biometeorology*, 62(3), 319–330.

ASHRAE, 1997, ASHRAE Fundamentals, Chapter 6. *Psychrometrics*, American Society for Heating, Refrigerating and Air conditioning Engineers, Atlanta, GA: Tullie Circle.

ASTM (American Society for Testing and Materials), 2005a, ASTM F237–05 Standard test method for measuring the evaporative resistance of clothing using a sweating manikin, West Conshohocken, PA: ASTM.

ASTM (American Society for Testing and Materials), 2005b, ASTM F1291–05 Standard test method for measuring the thermal insulation of clothing using a heated manikin, West Conshohocken, PA: ASTM.

Baron, R., 1972, Aggression as a function of ambient temperature and prior anger arousal, *Journal of Personality and Social Psychology*, 21, 183–189.

Belding, H. S. and Hatch, T. F., 1955, Index for evaluating heat stress in terms of resulting physiological strain, *Heating Piping and Air Conditioning*, 27, 129–136.

Bernard, T. E., Caravello, V., Schwartz, S. W. and Ashley, C. D., 2008, WBGT clothing adjustment factors for four clothing ensembles and the effects of metabolic demands, *Journal of Occupational Environmental Hygiene*, 5(1), 1–5.

Bethea, D. and Parsons, K. C., 2002, *The Development of a Practical Heat Stress Methodology for Use in UK Industry*. HSE Books. Research report No. 008. London, UK: HMSO.

Birnbaum, R. R. and Crockford, G. W., 1978, Measurement of the clothing ventilation index, *Applied Ergonomics*, 9(4), 194–200.

Blazejczyk, K., Epstein, Y., Jendritzky, G., Staiger, H. and Tinz, B., 2012, Comparison of UTCI to selected thermal indices, *International Journal of Biometeorology*, 56(3), 515–535.

Bonjer, F. H., Davies, C. J. M., Lange Andersen, K., Sargeant, A. J. and Shepherd, R. J., 1981, Measurement of maximum aerobic power, in Weiner, J. S. and Lourie, J. A. (eds.), *Practical Human Biology*, London: Academic Press.

Borg, G. A., 1982, Psychophysical bases of perceived exertion, *Medicine and Science in Sports and Exercise*, 14(5), 377.

Borg, G., 1998, *Borg's Perceived Exertion and Pain Scales*, Champaign, IL: Human Kinetics.

Botsford, J. H., 1971, A wet globe thermometer for environmental heat measurement, *American Industrial Hygiene Journal*, 32, 1–10.

Bouskill, L. M., 1999, Clothing ventilation and human thermal response. *PhD Thesis*, Department of Human Sciences, Loughborough University.

Bouskill, L. M., Havenith, G., Kuklane, K., Parsons, K. C. and Withey, W. R., 2002, Relationship between clothing ventilation and thermal insulation, *American Industrial Hygiene Association Journal*, 63(3), 263–268.

Bröde, P., Fiala, D., Lemke, B. and Kjellstrom, T., 2018, Estimated work ability in warm outdoor environments depends on the chosen heat stress assessment metric, *International Journal of Biometeorology*, 62(3), 331–345.

Broede, B. and Kampmann, B., 2017, Q10 effect and thermal cardiac reactivity related to the interrelation between heart rate and oxygen consumption under heat stress, in *Proceedings of ICEE*, Kobe.

BS 7963, 2000, Ergonomics of the thermal environment: Guide to the assessment of heat strain in workers wearing personal protective equipment, London: BSI, British Standards Institution.

BS EN ISO 7243, 2017, Ergonomics of the thermal environment—Assessment of heat stress using the WBGT (wet bulb globe temperature) index, 3rd edn, Geneva: International.

Budd, G. M., 2008, Wet-bulb globe temperature (WBGT)—Its history and its limitations. *Journal of Science and Medicine in Sport*, 11, 20–32.

Casanueva, A., Kotlarski, S., Fischer, A., Schwierz, C., Lemke, B. and Kjellstrom, T., 2018, European heat stress to reach critical levels under climate change conditions, in *EGU General Assembly Conference Abstracts 20*, Austria, p. 14130.

Ciuha, U., Gliha, M., Pogačar, T., Bogataj, L. K., Nybo, L., Flouris, A., Kjellstrom, T. and Mekjavic, I., 2017, Heat shield project: The effect of a summer heat wave on the productivity in an automobile-parts manufacturing plant, in *Proceedings of ICEE*, Kobe.

Clark, R. P. and Edholm, O. G., 1985, *Man and His Thermal Environment*, London: Edward Arnold.

Collins, K. J., Crockford, G. W. and Weiner, J. S., 1965, Sweat gland training by drugs and thermal stress, *Archives of Environmental Health*, 11, 407–420.

Crockford, G. W. and Rosenblum, H. A., 1974, The measurement of clothing microclimate volumes, *Clothing Research Journal*, 2(3), 109–114.

d'Ambrosio Alfano, F. R., Malchaire, J., Palella, B. I., Riccio G, 2014, WBGT index revisited after 60 years of use, *Annals of Occupational Hygiene*, 58, 955–970.

Dorman, L. E. and Havenith, G., 2009, The effects of protective clothing on energy consumption during different activities, *European Journal of Applied Physiology*, 105(3), 463–470.

Douglas, C. G., 1911, A method for determining the total respiratory exchange in man. *Journal of Physiology*, 42, 17–18.

DuBois, D. and DuBois, E. F., 1916, A formula to estimate surface area if height and weight are known, *Archives of Internal Medicine*, 17, 863.

DuBois, E. F., 1937, *Lane Medical Lectures. The Mechanism of Heat Loss and Temperature Regulation*, Stanford, CA: Stanford University Press.

Durnin, J. V. G. A. and Passmore, R., 1967, *Energy, Work and Leisure*, London: Heinemann Educational.

Edholm, O. G. and Weiner, J. S., 1981, *The Principles and Practice of Human Physiology*, London: Academic Press.

Ellis, F. P., Smith, F. E. and Walters, J. D., 1972, Measurement of environmental warmth in SI units, *British Journal of Industrial Medicine*, 29, 361–377.

EN 563, 1994, *Temperatures of Touchable Surfaces, Ergonomics Data to Establish Temperature Limit Values for Hot Surfaces*, Brussels: CEN.

EN ISO 7243, 2017, Ergonomics of the thermal environment—Assessment of heat stress using the WBGT (wet bulb globe temperature) index, 3rd edn, Brussels: CEN.

Fanger, P. O., 1970, *Thermal Comfort*, Copenhagen: Danish Technical Press.

Fiala, D., 1998, Dynamic simulation of human heat transfer and thermal comfort. *PhD Thesis*, De Montfort University.

Fiala, D., Havenith, G., Bröde, P., Kampmann, B. and Jendritzky, G., 2012, UTCI-Fiala multi-node model of human heat transfer and temperature regulation, *International Journal of Biometeorology*, 56(3), 429–441.

Fiala, D. and Havenith, G., 2016, Modelling human heat transfer and temperature regulation, in Gefen, A. and Epstein, Y. (eds.), *The Mechanobiology and Mechanophysiology of Military-Related Injuries*, pp. 265–302, Cham: Springer International Publishing. doi:10.1007/8415_2015_183.

Flouris, F. A., Dinas, P. C., Ioannou, L. G., Nybo, L., Havenith, G., Kenny, G. P., Kjellstrom, T., 2018, Workers' health and productivity under occupational heat strain: a systematic review and meta-analysis, *Lancet Planet Health*, 2018, 2, e521–e531, www.thelancet.com/planetary-health.

Fox, R. H., Goldsmith, R., Hampton, I. F. G. and Lewis, H. E., 1964, The nature of the increasing sweating capacity produced by heat acclimatization, *Journal of Physiology*, 171, 368–376.

Gagge, A. P., 1937, A new physiological variable associated with sensible and insensible perspiration, *American Journal of Physiology*, 120, 277–287.

Gagge, A. P., Burton, A. C. and Bazett, H. C., 1941, A practical system of units for the description of the heat exchange of man with his thermal environment, *Science*, 94, 428–430.

Gagge, A. P., Stolwijk, J. A. J. and Nishi, Y., 1971, An effective temperature scale based on a single model of human physiological temperature response, *ASHRAE Transactions*, 77, 247–262.

Gagge, A. P., Nishi, Y. and Gonzalez, R. R., 1973, Standard effective temperature: A single temperature index of temperature sensation and thermal discomfort, in *Proceedings of the CIB Commission W45 (Human Requirements) Symposium, Building Research Station, 13–15 September 1972*, Watford: HMSO.

Gagge, A. P. and Nishi, Y., 1983, Heat exchange between the human skin surface and the ther- mal environment, in Lee, D. (ed.) *Handbook of Physiology: Reactions to Environmental Agents*, pp. 69–92, Bethesda, MD: American Physiology Society.

Gagge, A. P., Fobelets, A. P. and Bergland, L. G., 1986, A standard predictive index of human response to the thermal environment, *ASHRAE Transactions*, 92(1), paper 92, 2B.

Gagnon, D., Kenny, G. P., 2011 Sex modulates whole-body sudomotor thermosensitivity during exercise. *Journal of Physiology*, 589(Pt 24), 6205–6217.

Gao, C. K., Kuklane, K., Östergren, P. O. and Kjellstrom, T., 2018, Occupational heat stress assessment and protective strategies in the context of climate change. *International Journal of Biometeorology*, 62(3), 359–371.

Garzón-Villalba, X. P., Wu, Y., Ashley, C. D. and Bernard, T. E., 2017, Ability to discriminate between sustainable and unsustainable heat stress exposures—Part 1: WBGT exposure limits, *Annals of Work Exposures and Health*, 61(6), 611–620. doi:10.1093/annweh/wxx034.

Garzón-Villalba, X. P., Wu, Y., Ashley, C. D. and Bernard, T. E., 2018, Heat stress risk profiles for three non-woven coveralls, *Journal of Occupational and Environmental Hygiene*, 15(1), 80–85.

Givoni, B., 1963, A new method for evaluating industrial heat exposure and maximum permissible work load, Paper submitted to the *International Biometeoro Logical Congress in Paris, France*, September.

Givoni, B. and Goldman, R. F., 1971, Predicting metabolic energy cost, *Journal of Applied Physiology*, 30(3), 429–433.

Givoni, B. and Goldman, R. F., 1972, Predicting rectal temperature response to work, environment and clothing, *Journal of Applied Physiology*, 2(6), 812–822.

Givoni, B. and Goldman, R. F., 1973, Predicting heart rate response to work, environment and clothing, *Journal of Applied Physiology*, 34(2), 201–204.

Givoni, B., 1976, *Man, Climate and Architecture*, 2nd edn, London: Applied Science.

Goldman, R. F., 1988, Standards for human exposure to heat, in Mekjavic, I. B., Banister, E. W. and Morrison, J. B. (eds.), *Environmental Ergonomics*, pp. 99–136, London: Taylor & Francis.

Goldman, R. F., 2006, Thermal manikins, their origins and role, in Fan, J. T. (ed.), *Proceedings of the International Conference on Thermal Manikins and Modelling*, Hong Kong: Hong Kong Polytechnic University.

Goldman, R. F. and Kampmann, B., 2007, *Handbook on Clothing*, 2nd edn. NATO Research Study Group 7 on Bio-medical research aspects of military protective clothing, Brussels. www.lboro.ac.uk/microsites/lds/EEC/ICEE/textsearch/Handbook%20on%20 Clothing%20-%202nd%20Ed.pdf.

Hancock, P. A., 1982, Task categorization and the limits of human performance in extreme heat, *Aviation, Space and Environmental Medicine*, 53, 778–784.

Hanson, M. and Graveling, R. A., 1999, Development of a draft British Standard, The assessment of heat strain for workers wearing personal protective equipment, IOM Research Report TM/99/03, IOM, Edinburgh.

Haslam, R. A., 1987, An evaluation of models of human response to hot and cold environments. *PhD Thesis*, Loughborough University.

Haslam, R. A. and Parsons, K. C., 1988, Quantifying the effects of clothing for models of human response to the thermal environment, *Ergonomics*, 31(12), 1787–1806.

Haslam, R. A. and Parsons, K. C., 1989a, Models of human response to hot and cold environments, in *Human Modelling Group Final Report*, vols. 1 and 2, Farnborough: APRE.

Haslam, R. A. and Parsons, K. C., 1989b, Computer based models of human responses to the thermal environments: Are their predictions accurate enough for practical use? in Mercer, J. B. (ed.), *Thermal Physiology*, Amsterdam: Elsevier.

Havenith, G., Heus, R., and Lotens, W. A., 1990, Clothing ventilation, vapour resistance and permeability index: changes due to posture, movement and wind, *Ergonomics*, 33(8), 989–1005.

Hensel, H., 1981, *Thermoreception and Temperature Regulation*, London: Academic Press.

Hodder, S. G. and Parsons, K. C., 2001, Field trials in Seville, in automotive glazing: Task 2.4—Thermal comfort, Final Technical Report NO. BE963020, Brite/Euram, European Commission, Brussels.

Hodder, S. G., 2002, Thermal comfort in vehicles: The effect of solar radiation. *PhD Thesis*, Loughborough University.

Hodder, S. G. and Parsons, K. C., 2007, The effects of solar radiation on thermal comfort, *International Journal of Biometeorology*, 51(3), 233–250.

Houghton, F. C. and Yagloglou, C. P., 1923, Determining equal comfort lines, *Journal of ASHVE*, 29, 165–176.

Hubbard, R. W. and Armstrong, L. E., 1988, The heat illnesses: Biochemical, ultra structural and fluid electrolyte, in Pandolf, K. B., Sawka, M. N. and Gonzalez, R. R. (eds.), *Human Performance Physiology and Environmental Medicine at Terrestrial Extremes*, pp. 1–44, Indianapolis, IN: Brown and Benchmark.

Humphreys, M. A., 1972, Clothing and thermal comfort of secondary school children in summertime, in *Proceedings of CIB Commission W45 Symposium Thermal Comfort and Moderate Heat Stress*, Watford: HMSO.

Inoue, Y. and Kondo, N., 2017, Aging, sweating, and sweat gland function, in *Proceedings of ICEE*, Kobe.

ISO 7243, 2017, Ergonomics of the thermal environment—Assessment of heat stress using the WBGT (wet bulb globe temperature) index, 3rd edn, Geneva: International Organization for Standardization.

ISO 7726, 1998, Ergonomics of the thermal environment—Instruments for measuring physical quantities, 2nd edn, Geneva: International Organization for Standardization.

ISO 7730, 2005, Ergonomics of the thermal environment—Analytical determination and interpretation of thermal comfort using calculation of the PMV and PPD indices and local thermal comfort criteria, 3rd edn, Geneva: International Organization for Standardization.

ISO 7933, 1989, Hot environments—Analytical determination and interpretation of thermal stress using calculation of required sweat rate, Geneva: International Standards Organization.

ISO 7933, 2004, Ergonomics of the thermal environment—Analytical determination and interpretation of heat stress using calculation of the predicted heat strain, 2nd edn, Geneva: International Organization for Standardization.

ISO 8996, 2004, Ergonomics of the thermal environment—Determination of metabolic rate, 2nd edn, Geneva: International Organization for Standardization.

ISO 9886, 2004, Evaluation of thermal strain by physiological measurements, 2nd edn, Geneva: International Organization for Standardization.
ISO 9920, 2007, Estimation of thermal insulation and water vapour resistance of a clothing ensemble (see also Amended Version 2009), 2nd edn, Geneva: International Organization for Standardization.
ISO 9920, 2009, Estimation of thermal insulation and water vapour resistance of a clothing ensemble (Amended Version to 2008), 2nd edn, Geneva: International Organization for Standardization.
ISO 11079, 2007, Ergonomics of the thermal environment—Determination and interpretation of cold stress when using required clothing insulation (IREQ) and local cooling effects, 1st edn, Geneva: International Organization for Standardization.
ISO 12894, 2001, Ergonomics of the thermal environment—Medical supervision of individuals exposed to extreme hot or cold environments, 1st edn, Geneva: International Organization for Standardization.
ISO 13732-1, 2001, Ergonomics of the thermal environment—Methods for the assessment of human responses to contact with surfaces—Part 1: Hot surfaces, 1st edn, Geneva: International Organization for Standardization.
ISO 13732-2, 2001, Method for the assessment of human responses to contact with surfaces (ISO DTR 13732)—Part 2: Human contact with surfaces at moderate temperature, London: BSI.
ISO 13732-3, 2005, Ergonomics of the thermal environment—Methods for the assessment of human responses to contact with surfaces—Part 3: Cold surfaces, 1st edn, Geneva: International Organization for Standardization.
ISO 15265, 2004, Ergonomics of the thermal environment—Risk assessment strategy for the prevention of stress or discomfort in thermal working conditions, 1st edn, Geneva: International Organization for Standardization.
ISO 28802, 2012, Ergonomics of the physical environment—Assessment of environments by means of an environmental survey involving physical measurements of the environment and subjective responses of people, Geneva: International Organization for Standardization.
ISO/AWI TR 16595, 2015, Working practices for hot environments, Geneva: International Organization for Standardization.
ISO DIS 7933, 1987, Hot environments: analytical determination and interpretation of thermal stress using calculation of required sweat rate. Geneva: International Organization for Standardization.
ISO DIS 7933, 2018, Ergonomics of the thermal environment—Analytical determination and interpretation of heat stress using calculation of the predicted heat strain, Geneva: International Organization for Standardization.
ISO NP TR 23454-1, 2018, Human performance in physical environments: Part 1—A performance framework.
Jacklitsch, B., Jon, W. W., Musolin, K., Coca, A., Kim, J.-H. and Turner, N., 2016, Criteria for a recommended standard. Occupational exposure to heat and hot environments, Washington DC: DHSS, National Institute for Occupational Safety and Health (NIOSH).
Jendritzky, G. and de Dear, R., (eds.), 2012, Special issue, universal thermal climate index (UTCI), *International Journal of Biometeorology*, 56(3), 419.
Jendritzky, G., de Dear, R. and Haventih, G., 2012, UTCI—Why another thermal index? *International Journal of Biometeorology*, 56(3), 421–428.

Kampmann, B. and Piekarski, C., 2000, The evaluation of workplaces subjected to heat loss: Can ISO 7933 (1989) adequately describe heat strain in industrial workplaces? *Applied Ergonomics*, 31(1), 59–72.

Kenney, W. L., Lewis, D. A., Hyde, D. E., Dyksterhouse, T. S., Armstrong, C. G., Fowler, S. R., Williams, D. A., 1988, Physiologically derived critical evaporative coefficients for protective clothing ensembles. *Journal of Applied Physiology*, 63, 1095–1099.

Kenney, W. L., Mikita, D. J., Havenith, G., Puhal, S. M. and Crosby, P., 1993, Simultaneous derivation of clothing specific heat exchange coefficients, *Medicine and Science in Sports and Exercise, Special Communications*, 25(2), 283–289.

Kenny, G.P., Yardley, J., Brown, C., Sigal, R.J., Jay, O., 2010, Heat stress in older individuals and patients with common chronic diseases. *CMAJ*, 182(10):1053–1060.

Kenney, L. W., Wilmore, J. H. and Costill, D. L., 2012, *Physiology in Sport and Exercise*, 5th edn, Champaign, IL: Human Kinetics.

Kenny, G., 2017, Environmental factors impacting responses of older individuals to heat, in *Proceedings of ICEE*, Kobe.

Kenshalo, D. R., 1970, Psychophysical studies of temperature sensitivity, in Neff, W. D. (ed.), *Contributions to Sensory Physiology*, New York: Academic Press.

Kerslake, D. M., 1972, *The Stress of Hot Environment*, Cambridge: Cambridge University Press.

Kilbourne, E. M., 1997, Heat waves and hot environments, in Noji, E. (ed.), *The Public Health Consequences of Disasters*, pp. 245–269, New York: Oxford University Press.

Kjellstrom, T., Gabrysch, S., Lemke, B., and Dear, K., 2009, The `Hothaps' programme for assessing climate change impacts on occupational health and productivity: an invitation to carry out field studies, *Global Health Action*, 2, doi: 10.3402/gha.v2i0.2082.

Kjellstrom, T., Sawada, S. -I., Bernard, T., Parsons, K. C., Rintamäki, H. and Holmér, I., 2013, Climate change and occupational heat problems, *Industrial Health*, 51(1), 1–2.

Kjellstrom, T., Freyberg, C., Lemke, B., Otto, M. and Briggs, D., 2018, Estimating population heat exposure and impacts on working people in conjunction with climate change *International Journal of Biometeorology*, 62(3), 291–306.

Kwon, J. Y., 2009, Human responses to outdoor thermal environments. *PhD Thesis*, Department of Human Sciences, Loughborough University.

Kwon, J. Y. and Parsons, K. C., 2009, Evaluation of the PMV thermal comfort index in outdoor weather conditions, *Journal of Energy and Building*, 35, 310–315.

Ladell, W. S. S., 1964, Terrestrial animals in humid heat: Man, in Dill, D. B. (ed.), *Handbook of Physiology, Section 4*, pp. 625–659, Washington, DC: American Physiology Society.

Legg, S. J. and Pateman, C. M., 1984, A physiological study of the repetitive lifting capabilities of healthy young males, *Ergonomics*, 27(3), 259–272.

Leithead, C. S. and Lind, A. R., 1964, *Heat Stress and Heat Disorders*, London: Cassell.

Lemke, B., Otto, M., Lines, L. and Kjellstrom, T., 2018, Geografic impact of climate change on peoples' work capacity, in *ISEE Conference Abstracts*, Ottawa, Canada.

Levi, M., Kjellstrom, T. and Baldasseroni, A., 2018, Impact of climate change on occupational and health and productivity: A systematic literature review focusing on workplace heat, *La Medicina del Lavoro*, 109(3), 163–179. doi: 10.23749/mdl.v109i3.6851.

Lewis, W. K., 1922, The evaporation of a liquid into gas, *ASME Transactions*, 44, 325–335.

Liddell, D. K., 1963, Estimation of energy expenditure from expired air, *Journal of Applied Physiology*, 18, 25–29.

Lind, A. R., 1963, A physiological criterion for setting thermal environmental limits for everybody's work, *Journal of Applied Physiology*, 18, 51–56.

Lind, A. R. and Bass, D. E., 1963, The optimal exposure time for the development of acclimatisation to heat, *Federal Proceedings*, 22, 704–708.

Lotens, W. A. and Havenith, G., 1988, Ventilation of rainwear determined by a trace gas method, in Mekjavic, I. B., Bannister, E. W. and Morrison, J. B. (eds.), *Environmental Ergonomics*, pp. 162–176, London: Taylor & Francis.

Lotens, W. A., 1993, Heat transfer from humans wearing clothing. *PhD Thesis*, University of Delft.

Lu, Y., Ma, N., Wang, L. and Wang, S., 2017, The impact of shape memory alloy size on the protective performance of fabrics used in fire-fighter's protective clothing, in *Proceedings of ICEE*, Kobe.

Machle, W. and Hatch, T. F., 1947, Heat: Mass exchanges and physiological responses. *Physiological Reviews*, 27, 200–227.

Mackworth, N. H., 1950, Researches on the measurement of human performance, Medical Research Council Special Report, No. 268, London: HMSO.

Malchaire, J., Gebhardt, H. J. and Piette, A., 1999, Strategy for evaluation and prevention of risk due to work in thermal environments. *The Annals of Occupational Hygiene*, 43(5), 367–376.

Malchaire, J., Piette, A., Kampmann, B., Havenith, G., Mehnert, P., Holmér, I., Gebhardt, H., Griefahn, B., Alfano, G. and Parsons, K. C., 2000, Development and validation of the predictive heat strain (PHS) model, in Werner, J. and Hexamer, M. (eds.), *Environmental Ergonomics IX*, pp. 133–136, Aachen: Shaker Verlag.

Malchaire, J., Piette, A., Kampmann, B., Mehnert, P., Gebhardt, H., Havenith, G., de Hartog, E., et al., 2001, Development and validation of the predicted heat strain model, *Annals of Occupational Hygiene*, 45(2), 123–135.

Maley, M., Minett, G., Bach, A., Zietek, S. and Stewart, I., 2017, Evaluation of commercial cooling systems to minimise thermal strain while wearing chemical-biological protective clothing, in *Proceedings of ICEE*, Kobe.

Marzetta, L. A., 1974, Engineering and construction manual for an instrument to make burn hazard measurements in consumer products, US Department of Commerce, National Bureau of Standards Technical Note 816, Washington DC.

Matthies, F., Bickler, G., Cardeñosa Marín, N. and Hales, S., 2008, *Heat—Health Action Plans*, Geneva: World Health Organization.

McArdle, B., Dunham, W., Holling, H. E., Ladell, W. S. S., Scott, J. W., Thomson, M. L. and Weiner, J. S., 1947, The prediction of the physiological effects of warm and hot environments, Medical Research Council, London, RNP Rep. 47/391.

McCullough, E. A., Jones, B. W. and Huck, J., 1985, A comprehensive database for estimating clothing insulation, *ASHRAE Transactions*, 91(2A), 29–47.

McCullough, E. A., Jones, B. W. and Tamura, T., 1989, A database for determining the evaporative resistance of clothing, *ASHRAE Transactions*, 95, 316–328.
McIntyre, D. A., 1980, *Indoor Climate*, London: Applied Science.
McNeill, M. and Parsons, K. C., 1996, Heat stress in nightclubs, in Robertson, S. A. (ed.), *Contemporary Ergonomics*, pp. 208–213, London: Taylor & Francis.
Meese, G. B., Kok, R., Lewis, M. I. and Wyan, D. P., 1984, A laboratory study of the effects of moderate thermal stress on the performance of factory workers, *Ergonomics*, 27(1), 19–43.
Metz, B., 1988, *Proceedings of CEC Seminar on Heat Stress Indices*, Luxembourg: Commission of the European Communities.
Millington, P. F. and Wilkinson, R., 2009, Skin, digital reprint of 1983 version, Cambridge University Press.
Montagna, W. and Parakkal, P. F., 1974, *The Structure and Function of Human Skin*, New York: Academic Press.
Monteith, J. L. and Unsworth, M. H., 1990, *Principles of Environmental Physics*, 2nd edn, London: Edward Arnold.
Moritz, A. R. and Henriques, F. C., 1947, Studies in thermal injury II. The relative importance of time and air surface temperatures in the causation of cutaneous burns, *American Journal of Pathology*, 23, 695–720.
Morrissey, S. J. and Liou, Y. H., 1984, Metabolic cost of load carriage with different container sizes, *Ergonomics*, 27(8), 847–853.
Murgatroyd, P. R., Shetty, P. S. and Prentice, A. M., 1993, Techniques for the measurement of human energy expenditure: A practical guide, *International Journal of Obesity*, 17, 549–568.
Nadel, E. R., Mitchell, J. W. and Stolwijk, J. A. J., 1973, Differential thermal sensitivity in the human skin, *Pflugers Archives*, 340, 71–76.
Neal, M. S., 1998, Development and application of a clothed thermoregulatory model. *PhD Thesis*, Department of Human Sciences, Loughborough University.
Nielsen, M., 1938, Die Regulation der Kopertemperatur bei Muskelerbeit, *Skandinavisches Archiv für Physiologie*, 79, 193–230.
Nielsen, B. and Nielsen, M., 1965, Body temperature during work at different environmental temperatures. *Acta Physiologica Scandinavica*, 64, 323–331.
NIOSH, 1986, Occupational exposure to hot environments, DHHS National Institute for Occupational Safety and Health (NIOSH), Washington, DC.
NIOSH, 2013, Workplace solutions: preventing heat-related illness or death of outdoor workers. Cincinnati, OH: U.S. Department of Health and Human Services, Centers for Disease Control and Prevention, National Institute for Occupational Safety and Health. Publication No. 2013-143.
NIOSH, 2016, Criteria for a recommended standard. Occupational exposure to heat and hot environments, DHSS, National Institute for Occupational Safety and Health, (NIOSH), Washington, DC.
Nishi, Y. and Gagge, A. P., 1977, Effective temperature scale useful for hypo and hyperbaric environments, *Aviation Space and Environmental Medicine*, 48, 97–107.
O'Brien, N. V., Parsons, K. C. and Lamont, D. R., 1997, Assessment of heat strain on workers in tunnels using compressed air compared to those in free air conditions, in Mining and Metallurgy, Tunnelling '97 Conference, pp. 341–352, London: The Institution of Mining and Metallurgy.

Olesen, B. W. and Dukes-Dubos, F. N., 1988, International standards for assessing the effect of clothing on heat tolerance and comfort, in Mansdorf, S. Z., Sager, R. and Nielson, A. P. (eds.), *Performance of Protective Clothing*, pp. 17–30, Philadelphia, PA: ASTM.

OSHA, 2011, OSHA-NIOSH infosheet: protecting workers from heat illness. Cincinnati, OH: U.S. Department of Health and Human Services, Centers for Disease Control and Prevention, National Institute for Occupational Safety and Health. DHHS (NIOSH) Publication No. 2011-174, http://www.cdc.gov/niosh/docs/2011-174/.

Pandolf, K. B., Givoni, B. and Goldman, R. F., 1977, Predicting energy expenditure with loads while standing or walking very slowly, *Journal of Applied Physiology*, 43(4), 577–581.

Parsons. K. C. and Clark, N., 1984, Evaluation of the PMV thermal comfort index, in Megaw, E. D. (ed), *Contemporary Ergonomics*, London: Taylor and Francis.

Parsons, K. C. and Hamley, E. L., 1989, Practical methods for the estimation of human metabolic heat production, in Mercer, J. B. (ed.), *Thermal Physiology*, pp. 777–781, Oxford: Excerpta Medica.

Parsons, K. C. and Bishop, D., 1991, A data base model of human responses to thermal environments, in Lovesey, E. J. (ed.), *Contemporary Ergonomics*, pp. 444–449, London: Taylor & Francis.

Parsons, K. C., 1992, The thermal audit, in Lovesey, E. J. (ed.), *Contemporary Ergonomics*, pp. 85–90, London: Taylor & Francis.

Parsons, K. C., 1993, *Human Thermal Environments*, 1st edn, New York: Taylor and Francis.

Parsons, K. C., 2000, An adaptive approach to the assessment of risk for workers wearing protective clothing in hot environments, in Kuklane, K. and Holmér, I. (eds), *Ergonomics of Protective Clothing*, pp. 34–37. Solna: National Institute for Working Life, Sweden.

Parsons, K. C., 2003, *Human Thermal Environments*, 2nd edn, New York: Taylor and Francis.

Parsons, K. C., 2005, Chapter 22, The environmental ergonomics survey, in Wilson, J. and Corlett, N. (eds.), *Evaluation of Human Work*, pp. 633–643, 3rd edn, New York: Taylor and Francis.

Parsons, K. C., 2006, Heat stress standard ISO7243 and its global application, *Industrial Health*, 44(3), 368–379.

Parsons, K. C., 2014, *Human Thermal Environments*, 3rd edn, New York: Taylor and Francis.

Parsons, K. C., 2018, ISO standards on physical environments for worker performance and productivity, *Industrial Health, 56, 93–95.*

Pennes, H. H., 1948, Analysis of tissue and arterial blood temperatures in the resting human forearm, *Journal of Applied Physiology*, 1, 93–122.

Pepler, R. D., 1964, Psychological effects of heat, Chapter 12, in Leithead, C. S. and Lind, R. (eds.), *Heat Stress and Heat Disorders*, pp. 237–253, London: Cassell.

PHE, 2018, Heatwave plan for England: Protecting health and reducing harm from severe heat and heatwave, Public Health England.

Pierce, F. T. and Rees, W. H., 1946, The transmission of heat through textile fabrics, Part II, *Journal of Textile Institute*, 37, 181–204.

Raccuglia, M. and Havenith, G., 2017, Wetness sensation during rest and exercise and the interaction with textiles: How can you feel it? in *Proceedings of ICEE*, Kobe.

Racinais S, Alonso J M, Coutts A J, Flouris A D, Girard O, 2015, Consensus recommendations on training and competing in the heat, *Scandinavian Journal of Medicine & Sports*, 25, 6–19.

Ramanathan, N., 1964, A new weighting system for mean surface temperature of the human body, *Journal of Applied Physiology*, 19(3), 531–533.

Ramsey, J. D., Burford, C. L., Beshir, M. Y. and Jensen, R. C., 1983, Effects of workplace thermal conditions of safe work behaviour, *Journal of Safety Research*, 14, 105–114.

Ramsey, J. D. and Kwon, Y. C., 1988, Simplified decision rules for predicting performance loss in the heat, in *Proceedings of a Seminar on Heat Stress Indices*, pp. 337–372, Luxembourg: Commission of the European Communities.

Randle, I. P. M., 1987, Predicting the metabolic cost of intermittent load carriage in the arms, in Megaw, E. D. (ed.), *Contemporary Ergonomics*, pp. 286–291, London: Taylor & Francis.

Renbourn, E. T. (ed.), 1972, *Materials and Clothing in Health and Disease*, London: H.K. Lewis.

Rohles, F. H. and Nevins, R. G., 1971, The nature of thermal comfort for sedentary man, *ASHRAE Transactions*, 77(1), 239–246.

Santee, W. R. and Gonzalez, R. R., 1988, Characteristics of the thermal environment, in Pandolf, K. B., Sawka, M. N. and Gonzalez, R. R. (eds.), *Human Performance Physiology and Environmental Medicine at Terrestrial Extremes*, pp. 1–44, Indianapolis, IN: Brown and Benchmark.

Sawka, M. N., 1988, Body fluid responses and hypohydration during exercise heat stress, in Pandolf, K. B., Sawka, M. N. and Gonzalez, R. R. (eds.), *Human Performance Physiology and Environmental Medicine at Terrestrial Extremes*, pp. 227–266, Indianapolis, IN: Brown and Benchmark.

Sawka, M. N and Wenger, C. B., 1988, Physiological responses to acute exercise heat stress, in Pandolf, K. B., Sawka, M. N. and Gonzalez, R. R. (eds.), *Human Performance Physiology and Environmental Medicine at Terrestrial Extremes*, pp. 227–266, Indianapolis, IN: Brown and Benchmark.

Sevitt, S., 1949, Local blood flow changes in experimental burns, *Journal of Pathological Bacteriology*, 61, 427–442.

Shibolet, S., Fisher, S., Gilat, T., Bank, H. and Heller, H., 1962, Fibrinolosis and hemorrhages in fatal heaystroke, *New England Journal of Medicine*, 266, 169–173.

Siekmann, H., 1989, Determination of maximum temperatures that can be tolerated on contact with hot surfaces, *Applied Ergonomics*, 20(4), 313–317.

Siekmann, H., 1990, Recommended maximum temperatures of touchable surfaces, *Applied Ergonomics*, 20(1), 69–73.

Smith, C. J. and Havenith, G., 2011, Body mapping of sweating patterns in male athletes in mild exercise-induced hyperthermia. *European Journal of Applied Physiology*, 111, 1391–1404.

Stolwijk, J. A. J. and Hardy, J. D., 1966, Temperature regulation in man—A theoretical study, *Pfluger Archives*, 291, 129–162.

Stolwijk, J. A. J. and Hardy, J. D., 1977, Control of body temperature, in *Handbook of Physiology, Section 9: Reaction to Environmental Agents*, Chapter 4, pp. 45–68, Bethesda, MD: American Physiological Society.

Tanabe, S. I., Kobayashi, K., Nakano, J., Ozeki, Y. and Konishi, M., 2002, Evaluation of thermal comfort using combined multinode thermoregulation (65MN) and radiation models and computational fluid dynamics (CFD), *Energy and Buildings*, 34(6), 637–646.

Teitlebaum, A. and Goldman, R. F., 1972, Increased energy cost with multiple clothing layers, *Journal of Applied Physiology*, 32(6), 743–744.

Turk, J., 1974, Development of a practical method of heat acclimatization for the army, in Monteith, J. L. and Mount, L. E. (eds.), *Heat Loss from Animal and Man*, London: Butterworth.

Underwood, C. R. and Ward, E. J., 1966, The solar radiation area of man, *Ergonomics*, 10, 399–410.

Vernon, H. M., 1919a, The influence of hours of work and of ventilation on output in tinplate manufacture, Report to Industrial Fatigue Research Board, No. 1, London: HMSO.

Vernon, H. M., 1919b, An investigation of the factors concerned in the causation of industrial accidents, Health of Munitions Workers Committee Memo No. 21, CD 9046.

Vernon, H. M., 1920, Fatigue and efficiency in the iron and steel industry, Report to Industrial Fatigue Research Board, No. 5, London: HMSO.

Vernon, H. M., Bedford, T. and Warner, C. G., 1927, The relations of atmospheric conditions to the working capacity and the accident rate of miners, Report to Industrial Fatigue Board, No. 39, London: HMSO.

Vernon, H. M., 1930, The measurement of radiant heat in relation to human comfort, *Journal of Physiology*, 70, 15.

Vernon, H. M., 1932, The measurement of radiant heat in relation to human comfort, Journal of Industrial Hygiene and Toxicology, 14, 95–111.

Vernon, H. M. and Warner, C. G., 1932, The influence of the humidity of the air on capacity for work at high temperatures, *Journal Hygiene Cambridge*, 32, 431–462.

Vogt, J. J., Candas, V., Libert, J. P. and Daull, F., 1981, Required sweat rate as an index of thermal strain in industry, in Cena, K. and Clark, J. A. (eds.), *Bioengineering, Thermal Physiology and Comfort*, pp. 99–110, Amsterdam: Elsevier.

Waldby, C., 2000, *The Visible Human Project: Informatic Bodies and Posthuman Medicine*, p. 4, London: Routledge.

Webb, L. H. and Parsons, K. C., 1997, Thermal comfort requirements for people with physical disabilities, in *Proceedings of the BEPAC and EPSRC Mini Conference: Sustainable Buildings*, pp. 114–121, Oxford: Abingdon.

Weir, J. B. and de, V., 1949, New methods for calculating metabolic rate with special reference to protein metabolism, *Journal of Physiology*, 109, 1–9.

Wenger, B. C., 1988, Human heat acclimatization, in Pandolf, K. B., Sawka, M. N. and Gonzalez, R. R. (eds.), *Human Performance Physiology and Environmental Medicine at Terrestrial Extremes*, pp. 227–266, Indianapolis, IN: Brown and Benchmark.

WHO, 1969, Health factors involved in working under conditions of heat stress, Technical Report 412, Geneva.

WHO, 2012, Public health advice on preventing health effects of heat. New and updated information for different audiences, Copenhagen: WHO.

Wing, J. F., 1965, A review of the effects of high ambient temperature on mental performance, Aerospace Medical Research Laboratories, AMRL-TR-65 102, (NTIS AD 624144).

Wissler, E. H., 1961, Steady state temperature distribution in man, *Journal of Applied Physiology*, 16, 734–740.

Wissler, E. H., 1985, Mathematical simulation of human thermal behavior using whole body models, in Shitzer, A. and Eberhart, R. C. (eds.), *Heat Transfer in Medicine and Biology*, vol. 1, pp. 325–373, New York, NY: Plenum Institute of Environmental Medicine.

Wood, E. J. and Bladon, P. T., 1985, The human skin, in *Studies in Biology*, vol. 164, London: Edward Arnold.

Woodcock, A. H., 1962, Moisture transfer in textile systems, *Textile Research Journal*, 8, 628–633.

Wyndham, C. H. and Strydom, N. B., 1969, Acclimatising man to heat in climatic rooms or mines, *Journal of South African Institute of Mining and Metallurgy*, 69, 60–64.

Wyon, D. P., 1970, Studies of children under imposed noise and heat stress, *Ergonomics*, 13(5), 598–612.

Yaglou, C. P. and Minard, D., 1956, Habitability studies in climatic extremes. Annual Report No. 9 to the Commission on Environmental Hygiene, AFEB, 26 February.

Yaglou, C. P. and Minard, D., 1957, Control of heat casualties at military training centers, *American Medical Association Archives of Industrial Health*, 16, 302–316.

Zhu, F., 2001, A human thermoregulatory model for predicting thermal sensation in transient building environments, *MSc Thesis*, Department of Architecture, University of Cambridge.

Index

A

B

C

S

T

Printed in the United States
by Baker & Taylor Publisher Services

Printed in the United States
by Baker & Taylor Publisher Services